MECHANISMS IN SCIENCE

In recent years what has come to be called the 'New Mechanism' has emerged as a framework for thinking about the philosophical assumptions underlying many areas of science, especially in sciences such as biology, neuroscience, and psychology. This book offers a fresh look at the role of mechanisms, by situating novel analyses of central philosophical issues related to mechanisms within a rich historical perspective of the concept of mechanism as well as detailed case studies of biological mechanisms (such as apoptosis). It develops a new position, Methodological Mechanism, according to which mechanisms are to be viewed as causal pathways that are theoretically described and are underpinned by networks of difference-making relations. In contrast to metaphysically inflated accounts, this study characterises mechanism as a concept-in-use in science that is deflationary and metaphysically neutral, but still methodologically useful and central to scientific practice.

STAVROS IOANNIDIS is Assistant Professor of Philosophy of Natural Sciences at the National and Kapodistrian University of Athens. He is principal investigator of the project MECHANISM, funded by the Hellenic Foundation for Research and Innovation.

STATHIS PSILLOS is Professor of Philosophy of Science and Metaphysics at the National and Kapodistrian University of Athens. He is the author of *Scientific Realism: How Science Tracks Truth* (1999) and *Causation and Explanation* (2002), and editor (with Henrik Lagerlund and Ben Hill) of *Reconsidering Causal Powers* (2021).

MECHANISMS IN SCIENCE

Method or Metaphysics?

STAVROS IOANNIDIS
University of Athens

STATHIS PSILLOS
University of Athens

CAMBRIDGE
UNIVERSITY PRESS

CAMBRIDGE
UNIVERSITY PRESS

University Printing House, Cambridge CB2 8BS, United Kingdom

One Liberty Plaza, 20th Floor, New York, NY 10006, USA

477 Williamstown Road, Port Melbourne, VIC 3207, Australia

314–321, 3rd Floor, Plot 3, Splendor Forum, Jasola District Centre, New Delhi – 110025, India

103 Penang Road, #05–06/07, Visioncrest Commercial, Singapore 238467

Cambridge University Press is part of the University of Cambridge.

It furthers the University's mission by disseminating knowledge in the pursuit of education, learning, and research at the highest international levels of excellence.

www.cambridge.org
Information on this title: www.cambridge.org/9781316519905
DOI: 10.1017/9781009019668

First published 2022

A catalogue record for this publication is available from the British Library.

Library of Congress Cataloging-in-Publication Data
NAMES: Ioannidis, Stavros, author. | Psillos, Stathis, 1965– author. Title: Mechanisms in science: method or metaphysics? / Stavros Ioannidis, University of Athens, Greece, Stathis Psillos, University of Athens, Greece.
DESCRIPTION: Cambridge, United Kingdom; New York, NY, USA: Cambridge University Press, 2022. | Includes bibliographical references and index.
IDENTIFIERS: LCCN 2021056613 (print) | LCCN 2021056614 (ebook) | ISBN 9781316519905 (hardback) | ISBN 9781009011495 (paperback) | ISBN 9781009019668 (epub)
SUBJECTS: LCSH: Mechanism (Philosophy) | Biology–Philosophy. Classification: LCC BD553 .I53 2022 (print) | LCC BD553 (ebook) | DDC 146/.6–dc23/eng/20211202
LC record available at https://lccn.loc.gov/2021056613
LC ebook record available at https://lccn.loc.gov/2021056614

ISBN 978-1-316-51990-5 Hardback

To my parents, Panagiota and Giorgos, with gratitude. —S.I.
To the memory of those two who made this exciting journey possible for
me, my father Dimitris (1927–2001) and my mother Maria
(1934–2019). —S.P.

Contents

Figures

Preface

This book is the product of genuinely collaborative work (the names appear in alphabetical order). We have known each other for almost two decades now. We met for the first time when S.I., still at high school, walked into S.P.'s office to ask advice concerning degrees in philosophy. S.I. became an undergraduate student in S.P.'s dept; then he did graduate studies in the philosophy of biology in the University of Bristol and joined the History and Philosophy of Science Department of the University of Athens as a postdoc in 2012.

Over the years, we have developed a philosophical partnership that extends from issues in the history of philosophy (mostly during the seventeenth century) to issues in contemporary metaphysics of science. Key to this partnership is mutual respect and tolerance as well as a common philosophical outlook. Both of us agree that good philosophy should be conceptually clear and historically sensitive. In fact, it seems that each and every philosophical problem is better illuminated if it is subjected to rigorous conceptual dissection; yet when the various parts are synthesised again, treating them in their historical concreteness enhances our understanding of their trajectory in time and space.

The project that led to this book started, like all of our joint ventures, with discussions over coffee on an early Saturday morning in a café in the centre of Athens. At the beginning of 2015, one of us (S.P.) received a kind invitation from Phyllis Illari and Stuart Glennan to contribute a piece on mechanisms and counterfactuals to the *Routledge Handbook of Mechanisms*. S.P. invited S.I. to work with him on this project. We started as we always did, by drafting the table of contents and dividing up the writing. Dozens of meetings and some heated discussions later, we submitted the piece in August 2016.

At the time of the mechanism project, we were both involved in a long and thorough study of the relations between metaphysics and physics in the seventeenth century. We started with Descartes and moved on to

Newton and Leibniz, with seminars, reading groups and workshops. We focused in particular on the transition from an Aristotelian power-based ontology to the modern law-based account of the world in terms of matter in motion. These endeavours brought with them the question of the relation between the Old Mechanism of the seventeenth century and the New Mechanism of the twenty-first. In October 2016 S.P. presented this historical narrative at the annual conference of L'academie Internationale de Philosphie des Sciences in Dortmund. The conference was on mechanisms and was organised by Brigitte Falkenburg and Gregor Schiemann.

The New Mechanism started to become the focal point of our research. Searching for a mechanism to study in detail, we came across the case of apoptosis, a.k.a. programmed cell death. We soon realised that that is a very rich case of a mechanism and it became the subject of our study. The more we thought about apoptosis, the more it became a showcase of our own approach to mechanism. This approach, which we called Methodological Mechanism, was aired first at conferences, most notably at the conference on Mechanisms in Medicine at the University of Kent at Canterbury in July 2017, organised by the gurus of mechanisms in the United Kingdom (the members of the 'Evaluating Evidence in Medicine' project (https://blogs.kent.ac.uk/jonw/projects/evaluating-evidence-in-medicine/).

The reception was mixed. John Worrall, our good friend, had to leave right after S.P. delivered the paper and on his way out he whispered in S.P.'s ear, 'You are back on the straight and narrow; you've become a logical positivist.' The hosts (Jon Williamson and Phyllis Illari) were more critical, arguing that the position is too thin, while María Jiménez Buedo (a young talented philosopher from Madrid) was enthusiastic. The result was a kind invitation from her to UNED in Madrid in May 2018, and a couple of talks in her research group (together with Jesús Zamora, Mauricio Suarez and David Teira) on the notion of mechanism in biology and the social sciences. In the meantime, our paper on apoptosis and Methodological Mechanism had appeared in print in the journal *Axiomathes*. Our main thesis, that the concept of mechanism in use in the sciences is mainly or exclusively methodological and not metaphysical, started to acquire some traction.

At roughly the same time, S.P. had finished a book review of Glennan's *The New Mechanical Philosophy* for the *Australasian Journal of Philosophy*. This book came to us as manna from heaven. Stuart presented very eloquently and forcefully the view that we wanted to oppose – the metaphysics-first view, as it were – and made us think harder about how best to defend our own practice-first approach. In the end, the review grew longer and longer and only a small part of it appeared in the *Australasian Journal of Philosophy*. The rest of it, focused as it was on activities qua a new ontic category, was destined to go into another paper, which was invited by the journal *Teorema*. In this paper we took on the concept of activity and developed our own difference-making account of the workings of a mechanism.

At the beginning of 2019, we received a kind email from our good friend Orly Shenker inviting us to a star-studded workshop in Jerusalem on the levels of reality, towards the end of May 2019. We took this opportunity to sharpen our thoughts on the issue of levels of mechanisms, and in particular on the relation of constitution that is supposed to hold between a mechanism and other mechanisms as its parts. That is an idea we reject in favour of causation. At the end of the day only S.I. managed to go as S.P.'s mother fell terminally ill and passed away a couple of weeks later.

The last station of this journey was in Geneva in September 2019, at the EPSA 19 Conference, where S.I. presented work concerning our main thesis, that mechanisms are causal pathways described in theoretical language. A few months later, just before the COVID-19 pandemic and the first lockdown, a good chunk of manuscript was submitted to Hilary Gaskin and Cambridge University Press. We should thank Hilary for her patience, care and support throughout the occasionally very demanding and tiring health-related issues that we both faced during the period of the completion of the manuscript. Two anonymous readers for Cambridge University Press made wonderfully detailed, critical but positive comments on the first draft. Without them the book would have been philosophically poorer.

Philosophy is essentially a communal enterprise. The book has benefitted a lot from a number of individuals who cared enough to make oral and written comments, ask questions and pose various challenges. The list of all those we would like to thank wholeheartedly is a lot bigger than the list of those mentioned by name. Hence our deepest thanks go to all those who asked questions and made comments, but whose names we don't know. But also to Konstantina Antiochou, Ken Binmore, Diderik Batens, Craig Callender, Nancy Cartwright, Paul Churchland, Peter Clark, Lindley Darden, Mauro Dorato, Jan Faye, Alexander Gebharter, Mania Georgatou, Michel Ghins, Donald Gillies, Olav Gjelsvik, Alan Hajek, Haris Hatziioannou, Chris Hitchcock, Carl Hoefer, Ilhan Inan, Gürol Irzik, Philip Kargopoulos, Patricia Kitcher, Buket Korkut, Daniel Kostić,

Vassilis Livanios, Peter Machamer, Vincent Müller, Daniel Nolan, Panagiotis Oulis (RIP), Kostas Pagondiotis, Demetris Portides, Nils Roll-Hansen, Pavlos Silvestros, Mauricio Suárez, Javier Suárez, David Teira, Amália Tsakiri, Eric Watkins, Erik Weber and Jim Woodward. Special mention goes to the following for giving us venues to present our work (as well as constructive criticism): Brigitte Falkenburg, Stuart Glennan, Meir Hemmo, Phyllis Illari, Valeriano Iranzo, Maria Jimenes Buedo, Federica Russo, Orly Shenker, Gregor Schieman, Erik Weber and Jon Williamson. We are also thankful to our research group in Athens – the usual suspects – who patiently heard and relentlessly criticised various drafts of chapters of the book, to Marilina Smyrnaki for her help in compiling the bibliography, to Marios Ioannidis for preparing the illustrations and to Stephanie Sakson for her careful and valuable copy-editing work.

Over the years, ideas that eventually formed parts of the book have been presented in seminars at the University of California San Diego, Caltech, American College of Thessaloniki, Bogazici University, University of Oslo, University of Ghent, UNED University in Madrid, Western University, Aristotle University of Thessaloniki and the University of Cyprus; also at the fourth Athens–Pittsburgh conference on Proof and Demonstration in Science and Philosophy, in Delphi (June 2003); at the workshop of the Metaphysics in Science Group in Athens (June 2003); at the Symposium on Mechanisms in the Sciences, APA Central Division, Chicago (April 2006); at the Conference on Causality in the Sciences, University of Kent (September 2008); at the Conference on Mechanisms and Causality in Science, University of Kent (September 2009); at the Workshop on the Metaphysics of Science, University of Warsaw (January 2010); at the Symposium on the Metaphysics of Science, College de France (May 2012); at ISHPSSB 2015 (Montreal, UQAM); at the AIPS Conference on Mechanistic Explanations, in Dortmund (October 2016); at the 'Mechanisms in Medicine' workshop, Centre of Reasoning, University of Kent (July 2017); at the Conference on the Multi-Level Structure of Reality (Israel Institute for Advanced Studies, Hebrew University of Jerusalem, and University of Haifa, May 2019); and at the EPSA19, Geneva (September 2019). S.P. would like to thank wholeheartedly the three women in his life, who help him get his bearings: Athena, Demetra and Artemis.

S.I.'s work has received funding from the Hellenic Foundation for Research and Innovation (HFRI) and the General Secretariat for Research and Innovation (GSRI), under grant agreement no. 1968.

Parts of this book have been based on reworked and expanded material that first appeared in journals and books. We thank the various publishers and editors for permission to use material in the book. Specifically:

Ioannidis, S. & Psillos, S. (2017). In defense of methodological mechanism: the case of apoptosis. *Axiomathes* 27 601–19. Reprinted by permission from Springer Nature, Copyright © 2017. **(Chapters 3, 4 and 10)**

Ioannidis, S. & Psillos, S. (2018). Mechanisms, counterfactuals and laws. In S. Glennan and P. Illari, eds., *The Routledge Handbook of Mechanisms and Mechanical Philosophy*, 1st ed. New York: Routledge, pp. 144–56. Copyright © 2018 by Routledge. Reproduced by permission of Taylor & Francis Group. **(Chapter 5)**

Ioannidis, S. & Psillos, S. (2018). Mechanisms in practice: a methodological approach. *Journal of Evaluation in Clinical Practice* 24: 1177–83. Copyright © 2018, John Wiley & Sons. **(Chapters 3 and 4)**

Psillos, S. (2004). A glimpse of the secret connexion: harmonizing mechanisms with counterfactuals. *Perspectives on Science* 12: 288–319. Copyright © 2004 The Massachusetts Institute of Technology. **(Chapters 6 and 7)**

Psillos, S. (2011). The idea of mechanism. In P. Illari, F. Russo, and J. Williamson, eds., *Causality in the Sciences*. Oxford: Oxford University Press, 771–88. This material has been reproduced by permission of Oxford University Press [http://global.oup.com/academic]. **(Chapter 2)**

Psillos, S. (2015). Counterfactual reasoning, qualitative: philosophical aspects. In J. Wright, ed., *International Encyclopedia of the Social & Behavioral Sciences*, 2nd ed. Oxford: Elsevier, vol. 5, pp. 87–94. Copyright © 2015, with permission from Elsevier. **(Chapter 7)**

Psillos, S. (2019). Review of Stuart Glennan, *The New Mechanical Philosophy. Australasian Journal of Philosophy* 97: 621–4. Copyright © Australasian Association of Philosophy, reprinted by permission of Taylor & Francis Ltd, http://www.tandfonline.com on behalf of Australasian Association of Philosophy. **(Chapter 6)**

Psillos, S. & Ioannidis, S. (2019). Mechanisms, then and now: from metaphysics to practice. In B. Falkenburg and G. Schiemann, eds., *Mechanistic Explanations in Physics and Beyond*. European Studies in Philosophy of Science. Cham: Springer Nature, pp. 11–31.

Reprinted by permission from Springer Nature, Copyright © 2019. **(Chapter 1)**

Psillos, S. & Ioannidis, S. (2019). Mechanistic causation: difference-making is enough. *Teorema* 38, no. 3: 53–75. www.unioviedo.es/Teorema **(Chapters 4 and 6)**

Introduction

When we think about mechanisms there are two general issues we need to consider. The first is broadly epistemic and has to do with the understanding of nature that identifying and knowing mechanisms yields. The second is broadly metaphysical and has to do with the status of mechanisms as building blocks of nature (and in particular, as fundamental constituents of causation). These two issues can be brought together under a certain assumption, which has had a long historical pedigree, namely, that nature is fundamentally mechanical.

That's a thought that was introduced by René Descartes and was popularised by Pierre Gassendi and Robert Boyle. Indeed, Descartes referred to his own theory of nature as 'Mechanica' since it considers sizes, shapes and motions, adding that it's a true theory of the world (Letter to Fromondus, 3 October 1637). Boyle called 'mechanical philosophy' his own corpuscularian theory which was based on matter and motion, two principles more 'catholic and universal' than any others. Mechanical philosophy was both a metaphysics of nature and a scientific theory. The point of contact was in the assumption that all worldly phenomena were ultimately the product of the mechanical affections of tiny corpuscles. That's the content of the old mechanist view that nature is *mechanical*.

And yet this very assumption has had no concrete ahistorical conceptual content. Rather, its content has varied according to the dominant conception of nature that has characterised each epoch. Nor has it been the case that the very idea of mechanism has had a fixed and definite content. Even if in the seventeenth century and beyond, the idea of mechanism had something to do with matter in motion subject to mechanical laws, *current* conceptions of mechanism have only a very loose connection with this assumption.

What kinds of commitments does more recent talk of mechanisms imply? Until the 1980s, the dominant views about mechanisms in the philosophy of science had been metaphysical. Mechanism has been seen as a view about causation: mechanisms were taken to provide the missing link (David Hume's 'secret connexion') between the cause and the effect. 'Mechanism', on this approach, is the very causal connection, and has been described in various ways as mark transmission (Salmon 1984), persistence, transference or possession of a conserved quantity (Mackie 1974; Salmon 1997; Dowe 2000).

In the 1990s, the New Mechanical Philosophy emerged, which is a view about the causal structure of the world: the world we live in is a world of mechanisms. A mechanism, nowadays, is virtually *any* relatively stable arrangement of entities such that, by engaging in certain interactions, a function is performed or an effect is brought about. Take a very typical characterisation of mechanism by William Bechtel and Adele Abrahamsen (2005, 423):

> (M) A mechanism is a structure performing a function in virtue of its component parts, component operations, and their organisation. The orchestrated functioning of the mechanism is responsible for one or more phenomena.

On this conception, a mechanism is *any* structure that is identified as such (i.e., as possessing a certain causal unity) via the function it performs. Moreover, a mechanism is a *complex* entity whose behaviour (i.e., the function it performs) is determined by the properties, relations and inter-actions of its parts. This priority of the parts over the whole – and in particular, the view that the behaviour of the whole is determined by the behaviour of its parts – is the distinctive feature of this broad account of mechanism.

Behind this broad understanding of mechanism, there is a certain metaphysics of nature. New mechanists take 'mechanisms' to be complex *entities* in their own right which are characterised by a certain ontological signature. This is supposed to be a signature that all mechanisms share in common, something that unifies all mechanisms and grounds their role in causation and explanation. The dominant view is that mechanisms are *structured wholes of entities and activities*. The latter are supposed to be the 'ontic correlate' of verbs; they are meant to ground the productive relations among entities and the productivity of the mechanism as a whole.

This metaphysics of mechanisms, a.k.a. 'new mechanical ontology' of entities, activities, organisation of parts into wholes and so on, invites a number of questions: What, in general terms, are the constituents of mechanisms? And what are their relations with more traditional metaphys-ical categories, such as objects, properties, powers and processes? So, fundamental ontology is brought to bear on how best to understand mechanisms in science.

A main claim of this book is that those philosophical views that offer metaphysically 'inflated' accounts of mechanisms are not necessary in order to understand scientific practice. And not just that. We shall also claim that there is no argument from the practice of using mechanisms in science to any metaphysics of mechanisms. That's mainly because we take it that the concept of mechanism as it is used in science, and in biology in particular, is methodological. We call our view Methodological Mechanism (MM). The main tenet of MM is that commitment to mechanism in science is first and foremost a methodological stance. The core of MM is a deflationary account of mechanism that is ontologically non-committal. According to what we call *Causal Mechanism*,

(CM) A mechanism is a causal pathway that is described in theoretical language.

Moreover, we claim that commitment to mechanism in science means adopting a certain methodological postulate, that is, that one should always look for the causal pathways producing the phenomena of interest. As such, it does not make any general ontological assertions about the 'deep' nature of causal processes.

On our deflationary account, talk of mechanisms in (biological) practice is talk about how causes (described in the language of theories) operate to bring about a certain effect. To identify a mechanism, then, is to identify a specific causal pathway that connects an initial 'cause' (the causal agent) with a specific result. Wherever there is a cause for a specific effect, there exists a mechanism that accounts for how the cause operates. The scientific task, then, is to identify the *mode of operation* of the cause, that is, the causal pathway. Identification of the causal pathway is crucial in order to establish that a causal link exists between a putative causal agent and a result (e.g., a disease state). Moreover, knowing the causal pathway makes interventions possible (and in the case of pathology, treatment). Our examination and defence of CM will unravel some limitations inherent in any attempt to extract metaphysical conclusions from scientific practice.

Admittedly, CM is thin. But it does *not* follow from CM that mechanisms are not 'things in the world'. After all, they *are* causal pathways! In characterising CM as deflationary or metaphysically agnostic, the point is that there is no need to say something 'deeper' than this in order to have a useful concept of mechanism that elucidates practice: a mechanism simply *is* a sequence of causal steps (or a process) that leads from an initial 'cause' to an end result. In sum: mechanisms in science and, in particular, in biology are stable causal pathways, described in the language of theory, where to identify a *causal* pathway is to identify difference-making relations among its components. The best way to introduce the main thesis of the book is by a parable.

A Parable

Imagine you've been a lifelong metaphysician trying to understand the fundamental building blocks of reality. You've read (more or less) everything written on the subject, which after an accident of classification, came to be called 'metaphysics'. You started with Aristotle's treatises that deal with the fundamental ontological categories, and which because they were placed after his φυσικά (physica; physics), were collectively dubbed μετά τά φυσικά (metaphysica; metaphysics). And in the fullness of time, you acquired views about all important topics: universals versus particulars; categorical versus dispositional properties; necessitarian versus non-necessitarian laws of nature; Humean versus non-Humean accounts of causation; and so on. You've been particularly excited by the concept of mechanism. You read about the mechanical account of nature that emerged in the seventeenth century, you reflected on the alternative, non-mechanical, way of explanation. You came to believe that causation is intimately connected with the presence of mechanism. And of course you lived through the revival of the mechanistic account of nature in the end of the twentieth century.

Metaphysics being what it is, that is, quite remote from ordinary and scientific experience, you had various doubts about the theories you entertained but at no time did you doubt that there are well-founded answers to the questions you grappled with. Now (that's the beginning of the parable), you are standing at the Pearly Gates about to meet your maker. After the usual introductions (with lots of rhetorical questions on God's part, 'Did you enjoy what you did for a living?,' etc.) God asks you the question you've been waiting and longing for: 'My child, is there anything you'd like to ask me because you haven't been satisfied with the answers you've hit upon yourself?' You gather all the strength you have and say: 'Yes, my Lord, there are indeed quite a few things that gave me sleepless nights and I'd die to find out the answer.' God being very busy replies candidly that only three short questions are allowed. Here is your first: Are there universals? God replies. (Unfortunately, there is no record of his answer.) Here is your second: 'Are there mechanisms in nature?' God reflects for a minute and then replies: 'Certainly! Light reflection, chemical bonding, mitosis but also economic crises and demonstrations are mechanisms. Where there is causation, there is mechanism!' Time for the third question, God says. You are puzzled. You collect all of your philosophical might and ask: 'Well, that's a list! Isn't there anything all these, and plenty of others like those in the list, share in common in virtue of which they are *mechanisms*? Isn't there something metaphysically deep that constitutes a mechanism?' God is somewhat upset since as he pointedly remarks these are *two* questions, but, knowing that his interlocutor is a philosopher he let it pass. 'So', you say with genuine aporia in your eyes. 'Well', he says, 'they are all causal pathways described in some theoretical language. That's what they have in common.' And before he parts company he winks and adds: 'I'm afraid you were barking up the wrong tree. "Mechanism" is everywhere as a piece in methodology but not in ontology.'

The Aim and Scope of Our Argument

The very aim of the book at hand is to explain the foregoing reply in detail and to show its plausibility. Lest we are misunderstood, that's not God's reply (who knows what this might be). But it's a possible reply to be assessed as any other philosophical theory.

We should be clear from the outset about the scope of our argument. Let us again distinguish between two questions one may ask about causation, one metaphysical, the other methodological. The metaphysical one concerns how exactly causation bottoms out: Does it bottom out in irreducible productive activities, in the manifestation of powers, in Humean regularities or (perhaps) in primitive difference-making relations? The methodological question is this: If we focus on how scientists discover mechanisms and how experiments are being done, are difference-making relations enough in order to understand scientific practice, or do we need also to say something about the ultimate nature of the truth-makers, if there are any, of these difference-making relations?

Concerning the question about methodology, we claim that if we focus on methodology and the epistemology of practice, difference-making relations should be enough (the case of the cause of scurvy, discussed in Chapter 4, will drive this point home). We do not have to say anything more about the truth-makers of causal claims. So, if one just focuses on the methodological question, one can remain agnostic about the metaphysics. It is in this sense that we described CM as a metaphysically agnostic position.

Concerning the question of metaphysics, one can, surely, give reasons to prefer some particular metaphysical picture over others. Our point is that whatever else these reasons are, they are not related to scientific practice. That is, what we emphatically reject is a particular kind of strategy of offering arguments in favour of a specific metaphysics of causation: that in order to understand the concept of mechanism *as it is used in the sciences* we need to be committed to a layer of thick metaphysical facts, for example, about activities or powers, as the truth-makers of causal claims.

In fact, in the subsequent chapters we will offer reasons to favour a particular metaphysical picture, namely, a Humean view that grounds difference-making relations in laws of nature understood as regularities, and to reject activities-based views that several mechanists accept. What is important to note however is that the argument against activities and in favour of Humean regularities will not be based on scientific practice. Rather, it will be based on an examination of the philosophical merits of the various specific metaphysical views. This means that if one is not inclined to pose the metaphysical question about causation and wishes to confine oneself to what is licensed by scientific practice, one can adopt CM for the concept-in-use and remain a metaphysical agnostic.

Given our complete view that combines a practiced-based CM with a particular approach to metaphysics, two opposite reactions are possible if one wants to reject our position. One can either opt for a stronger account about the metaphysics of causation, or an even more minimal metaphysical view. Let us be clear about each of them (although a fuller answer will be given in subsequent chapters).

When it comes to the search for stronger accounts that ground difference-making relations to something metaphysically robust, we take the central issue to be whether these accounts can tell illuminating stories about how mechanisms are used in science – we think that they do not. The point can be made as follows: when it comes to the practice of discovering difference-making relations, it does not matter whether we live in a Humean or a non-Humean world containing perhaps irreducible activities; in all cases, scientific practice would be the same and the search for mechanisms equally prevalent. This means, again, that such stronger accounts of causation do not help us understand scientific practice – of course, as noted already, one may have other kinds of reasons to favour a Humean metaphysics, for example, on grounds of conceptual economy.

Accounts of causation that may assume difference-making relations while not requiring an explicit stance on laws rely typically on a notion of variation among the values of relevant variables or invariance under interventions, which is less robust than regularity. Our differences with such accounts are less important than the similarities. Hence, they dismiss calls for metaphysically thicker views of causation and focus on how causal statements are established empirically. The key problem with such accounts is to avoid collapsing causation to correlations.

Conceptual Geography

In the book, we draw a number of distinctions and offer a number of characterisations of mechanism. To avoid confusion, we shall try here to offer a map of the conceptual landscape of mechanism.

The central distinction we draw is between the Old Mechanism and the New Mechanism, which is based on two pillars. The first is historical, while the second is conceptual. Old Mechanism emerged as a reaction to Aristotelian natural philosophy; it was foremost a theory of the natural world backed up by a metaphysics of nature. It was an attempt to leave behind the physics and the metaphysics of later scholasticism and to replace it with a new physics. The relation of the new physics with metaphysics was a matter of dispute. The clearest relation was in Descartes's metaphysical physics, to use Dan Garber's apt expression. By contrast, New Mechanism is, by and large, a philosophical re-interpretation of scientific practice and not a new science. As such, it is mostly a new metaphysics of nature. The second pillar of the distinction concerns the key features of the new metaphysics of science. The key feature of the metaphysics of Old Mechanism was that everything material was explainable by reference to shape, size and motion. A mechanism was any configuration of matter in motion subject to laws. By contrast to this 'flat' ontology, New Mechanism takes *any* kind of thing to be a mechanism provided it has a certain structure of entities engaging in activities, the entities and the activities being a lot more 'colourful' than the 'dull' entities (corpuscles) and the 'activities' (forces acting by contact) of Old Mechanism. A bit provocatively put, a mechanism of the new mechanical metaphysics is a matter of structure whereas Old Mechanism is a matter of content.

Occasionally we refer to 'Old Mechanism (narrowly understood)' aiming to draw a distinction *within* Old Mechanism between a narrow (Cartesian) understanding of mechanism as involving contact-action and a more liberal (post-Newtonian) account of mechanism which takes it that a mechanism is any system subject to Newton's laws, namely, the laws of *mechanics*. As Robert Schofield (1970, 15) has argued, post-Newtonian mechanists took it that 'causation for all the phenomena of nature was ultimately to be sought in the primary particles of an undifferentiable matter, the various sizes and shapes of possible combinations of these particles, their motions, and the forces of attraction and repulsion between them which determine these motions'. By the time of Henri Poincaré, this more liberal account of Old Mechanism was the dominant one.

This more liberal version of Old Mechanism we call 'mechanical mechanism' and we contrast it with what we call following A. C. Ewing 'quasi-mechanical' mechanism. That's a distinction that can, arguably, be traced to Immanuel Kant's *Third Critique* and is meant to introduce a conception of mechanism such that the properties of the whole are determined by the properties of its parts, but with no particular reference to mechanical properties and laws. Rom Harré has called these mechanisms *generative mechanisms*. They are taken to underpin causal connections: in virtue of them causes are supposed to produce their effects. As Harré aptly put it: 'not all mechanisms are mechanical' (1972, 118). This idea of quasi-mechanism or generative mechanism is a precursor of New Mechanism.

Finally, we draw a distinction between mechanism-of and mechanism-for, which relates to the role of mechanisms in causation. The dominant conception among new mechanists is that a mechanism is *for* a behaviour or function. The function/behaviour of a mechanism determines the boundaries of the mechanism and the identification of its components and operations. For us, however, the notion of a mechanism-for captures the functional notion of mechanism. A mechanism-of is any causal process, irrespective of whether or not a function is performed. We will argue that every mechanism-for is a mechanism-of, but not conversely. The mechanism-for/-of distinction serves to illustrate the fact that some causal pathways have a function and a functional role to play (they are mechanisms-for) while other causal pathways do not (they are mechanisms-of). The distinction then is important for driving home the point that there are mechanisms everywhere (where there are causal pathways) even if there is no function they perform.

The Road Map

Having drawn the conceptual map of mechanism, it's time to move to an orderly summary of the chapters of the book.

In Part I we look at some main aspects of the historical development of the concept of mechanism. This broader narrative is motivated by the explicit intention of the new mechanists to link the current mechanical philosophy with its older counterpart in the seventeenth century.

Chapter 1 examines the relationship between Old and New Mechanism and uses it to illuminate the relations between metaphysical and methodological conceptions of mechanism. This historical examination will directly motivate our new deflationary account of mechanism developed in the subsequent chapters. We start by focusing on the role of mechanistic explanation in seventeenth-century scientific practice, by discussing the views of Descartes, Christiaan Huygens, Gottfried Wilhelm Leibniz and Boyle, and the attempted mechanical explanations of gravity by Descartes and Huygens. We thereby illustrate how the metaphysics of Old Mechanism constrained scientific explanation. We then turn our attention to Isaac Newton's critique of mechanism. The key point is that Newton introduces a new methodology that frees scientific explanation from the metaphysical constraints of the older mechanical philosophy. Last, we draw analogies between Newton's critique of Old Mechanism and our critique of New Mechanism. The main point is that causal explanation in the sciences is legitimate even if we bracket the issue of what mechanisms or causes are as things in the world.

In Chapter 2, we continue our historical discussion of mechanistic explanation. The chief purpose of this chapter is to disentangle what we call *mechanical* and *quasi-mechanical* mechanism and point to the key problems they face. We begin by offering an outline of the mechanical conception of mechanism, as this was developed after the seventeenth century. We then present Poincaré's critique of mechanical mechanism in relation to the principle of conservation of energy. The gist of this critique is that mechanical mechanisms are too easy to be informative, provided that energy is conserved. We then advance the quasi-mechanical conception of mechanism and reconstruct G. W. F. Hegel's critique of the idea of quasi-mechanism, as this was developed in his *Science of Logic*. Hegel's problem, in essence, was that the unity that mechanisms possess is external to them and that the very idea that *all* explanation is mechanical is devoid of content. Finally, we bring together Poincaré's problem and Hegel's problem and argue that though mechanisms are not the building blocks of nature, the search for mechanism is epistemologically and methodologically welcome.

Part II develops our own science-first difference-making account of Causal Mechanism.

In Chapter 3 we present the first main part of our case for CM, by discussing in detail apoptosis, a central biological mechanism. We examine how Kerr and his co-workers first introduced apoptosis in 1972. We then present the most important stages in scientific research regarding apoptosis during the last decades that led to its identification as a central biological mechanism, explaining the shift from morphological descriptions to biochemical descriptions of the mechanism. We generalise the molecular definition of a pathway to arrive at a more general notion of a causal pathway. We also show that several distinctions used by biologists in order to differentiate between causal pathways and identify the genuine biological mechanisms (active vs passive, programmed vs non-programmed, physiological vs accidental) do not correspond to internal features of causal pathways, but concern an external feature, that is, the role those processes play within the organism.

In Chapter 4 we build on this discussion in order to argue that understanding mechanisms in the CM sense is all that is needed in order to understand biological practice. We clarify the main commitments of our view by presenting three theses that together constitute CM. (1) Mechanisms are to be identified with causal pathways; (2) causal relations among the components of a pathway are to be viewed in terms of difference-making; and (3) CM is metaphysically agnostic. A key point is that, in contrast to mechanistic theories of causation, for CM causation as difference-making is conceptually prior to the notion of a mechanism. We examine in some detail the discovery of the mechanism of scurvy in order to argue that difference-making is what matters in practice. We then turn to the main inflationary accounts of mechanism and contrast them with our deflationary view and its metaphysical agnosticism. We argue that CM offers a general characterisation of mechanism as a concept-in-use in the life sciences that is deflationary and thin, but still methodologically important.

In Chapter 5, we examine the relation between mechanisms and laws/counterfactuals by revisiting the main notions of mechanism found in the literature. We distinguish between two different conceptions of 'mechanism': *mechanisms-of* and *mechanisms-for*. We argue that for both mechanisms-of and mechanisms-for, counterfactuals and laws are central for understanding within-mechanism interactions. Concerning mechanisms-for, we claim that the existence of irregular mechanisms is compatible with the view that mechanisms operate according to laws. The discussion in this chapter, then, points to an asymmetrical dependence between mechanisms and laws/counterfactuals: while some laws and counterfactuals must be taken as primitive (non-mechanistic) facts of the world, all mechanisms depend on laws/counterfactuals.

In Chapter 6, we defend the *difference-making* thesis of CM, that is, the view that mechanisms are underpinned by networks of difference-making relations, by showing that difference-making is more fundamental than production in understanding mechanistic causation. Our argument is two-fold. First, we criticise Stuart Glennan's claim that mechanisms can be viewed as the truth-makers of counterfactuals and argue that counterfactuals should be viewed as metaphysically more fundamental. Second, we argue against the view that the productivity of mechanisms requires thinking of them as involving activities, qua a different ontic category. We criticise two different routes to activities: Glennan's top-down approach and Phyllis Illari and Jon Williamson's bottom-up approach. Given these difficulties with activities and mechanistic production, it seems more promising to start with difference-making and give an account of mechanisms in terms of it.

Given the centrality of counterfactual difference-making relations to the argument of the book, in Chapter 7 we say a few more things about the contrary-to-fact conditionals. We offer a primer on the logic and the semantics of counterfactuals, focusing on the two main schools of thought: the metalinguistic and the possible-worlds approach. We also present and examine James Woodward's interventionist counterfactuals and the Rubin-Holland model. We argue that the counterfactual approach is more basic than the mechanistic, but information about mechanisms can help sort out some of the methodological problems faced by the counterfactual account.

In Part III we show how our account of Causal Mechanism goes beyond New Mechanism and defend our own Methodological Mechanism.

In Chapter 8 we examine Carl Craver's well-known account of constitutive mechanisms, which takes the organised entities and activities that are the components of the mechanism to constitute the phenomenon to be explained. The main aim of the chapter is to criticise the adequacy of this view for illuminating mechanism as a concept-in-use in biological practice. We identify two main problems for the constitutive view: the problem of external components and the fact that some mechanisms can exist outside the entity the behaviour of which they underlie; we argue that both problems undermine the usefulness and appropriateness of viewing typical and paradigmatic cases of biological mechanisms in constitutive terms. The main claim of the chapter is that in order to understand the notion of mechanism as a concept-in-use, there is no need to posit a non-causal relation of constitution.

In Chapter 9 we present and defend a causal account of multilevel mechanistic explanation by examining various case studies from biology. We argue that two key consequences of Causal Mechanism are: (1) that levels and mechanisms are distinct notions and (2) that levels of multilevel explanations are levels of composition. This view is in stark contrast to Craver's account according to which levels in multilevel explanations are levels of mechanisms and multilevel explanations are instances of constitutive explanations. A key claim of the chapter is that whatever contributes to the phenomenon is part of the same pathway; but causal pathways can contain entities from multiple levels of composition. In order to motivate and illustrate our view, we use various examples from biology and medicine. We criticise some common views associated with the picture of a hierarchy of mechanistic levels and argue that our view allows for causation at higher levels.

In Chapter 10 we pick up the various threads and defend the main thesis of the book (Methodological Mechanism), namely, the claim that to be committed to mechanism is to adopt a certain methodological postulate, that is, to look for causal pathways for the phenomena of interest. We compare our view of Methodological Mechanism with an important discussion by Joseph Henry Woodger (1929) concerning the meaning of mechanism, which has been ignored in current discussions, as well as with the views of Robert Brandon. We then formulate a dilemma that new mechanists face; the dilemma arises from the unstable combination of two main tenets of New Mechanism, an ontological and a methodological one, both of which depend on the general characterisation of mechanism but that pull in opposite directions. We argue that CM is able to resolve the dilemma, by providing the best defence of the methodological tenet of New Mechanism, while at the same time preventing the adoption of a robust version of the ontological tenet.

In the last chapter, the Finale, we examine to what extent CM can be seen as a descendant of the original notion of mechanism developed in seventeenth century, by examining possible extensions of the seventeenth-century notion of mechanism and discussing whether they can be used to characterise mechanism as a concept-in-use.

In sum: Methodological Mechanism is mechanism enough.

PART I

Ideas of Mechanism

CHAPTER I

Mechanisms, Then and Now

1.1 Preliminaries

Let's call 'Old Mechanism' (or mechanical natural philosophy) the general conception about the nature of the world and of science that was developed in the seventeenth century by thinkers such as Descartes, Boyle, Huygens and Leibniz. The currently popular New Mechanism too constitutes a general framework for understanding science and nature, which emerged and spread in philosophy of science in the beginning of the twenty-first century. In this chapter, we look at some main aspects of the historical development of the concept of mechanism, by examining the differences and similarities between New and Old Mechanism. The motivation comes from the explicit intention of new mechanists to link the current mechanical philosophy with its older counterpart in the seventeenth century (see Machamer et al. 2000). However, there are several questions to be asked about the real relationship between the new and old mechanical philosophy.

A key similarity between Old and New Mechanism is that, for many old and new mechanists, Mechanism is both a metaphysical position about the structure of reality and a methodological thesis about the general form of scientific explanation and the methodology of science. The main aim of this chapter is to examine the relation between the metaphysics of mechanisms and the methodological role of mechanical explanation in the practice of science, by presenting and comparing the key tenets of Old and New Mechanism. We will use this historical examination to motivate our deflationary account of mechanism developed in the book.

1.2 Old versus New Mechanism

The mechanical world view of the seventeenth century was both a metaphysical thesis and a scientific theory. It was a metaphysical thesis insofar as it was committed to a reductionist account of all worldly phenomena to configurations of matter in motion subject to laws. In particular, it was committed to the view that all macroscopic phenomena are caused, and hence are accounted for, by the interactions of invisible microscopic material corpuscles. Margaret Wilson captured this view succinctly:

> The mechanism characteristic of the new science of the seventeenth century may be briefly characterised as follows: Mechanists held that all macroscopic bodily phenomena result from the motions and impacts of submicroscopic particles, or corpuscles, each of which can be fully characterised in terms of a strictly limited range of (primary) properties: size, shape, motion and, perhaps, solidity and impenetrability. (1999, xiii–xiv)

But this metaphysical thesis did, at the same time, license a *scientific theory* of the world, namely, a certain conception of scientific explanation and of theory construction. To offer a scientific explanation of a worldly phenomenon X was to provide a configuration Y of matter in motion, subject to laws, such that Y could cause X. A mechanical explanation then was (a species of) *causal explanation*: to explain that Y causes X was tantamount to constructing a mechanical model of how Y brings about X. The model was mechanical insofar as it was based on resources licensed by the metaphysical world view, namely, action of particles by contact in virtue of their primary qualities and subject to laws of motion.[1]

Nearly four centuries later, the mechanical world view has become prominent again within philosophy of science. It's become known as the 'New Mechanical Philosophy' or the 'New Mechanism' and has similar aspirations as the old one. New Mechanism, as Stuart Glennan puts it, 'says of nature that most or all the phenomena found in nature depend on mechanisms – collections of entities whose activities and interactions, suitably organized, are responsible for these phenomena. It says of science that its chief business is the construction of models that describe, predict, and explain these mechanism-dependent phenomena' (2017, 1).

So, New Mechanism too is a view about both science *and* the metaphysics of nature. And yet, in New Mechanism the primary focus has been on scientific practice, and in particular on the use of mechanisms in discovery, reasoning and representation (see Glennan 2017, 12). The focus on the metaphysics of mechanisms has emerged as an attempt to draw conclusions about the ontic signature of the world starting from the

---

[1] For the purposes of this chapter, we ignore issues of mind-body causation and we focus on body-body causation. We also ignore divisions among mechanists concerning the nature of corpuscles, the existence of vacuums, etc.

concept of mechanism as it is used in the sciences. According to Glennan, as the research into the use of mechanism in science developed, 'it has been clear to many participants in the discussion that metaphysical questions are unavoidable' (2017, 12). It is fair to say that New Mechanism aims to ground the metaphysics of mechanisms on the practice of mechanical explanation in the sciences.

1.3 Old Mechanism: From Metaphysics to Practice

A rather typical example of the interplay between the metaphysical world view and the scientific conception of the world in the seventeenth century was the attempted mechanical explanation of gravity.

1.3.1 *Mechanical Models of Gravity*

Let us start with Descartes. The central aim of the third and fourth parts of Descartes's *Principia Philosophiae*, published in 1644, was the construction of an account of natural phenomena. In Cartesian physics, the possible empirical models of the world are restricted from above by a priori principles which capture the fundamental laws of nature and from below by experience. Between these two levels there are various theoretical hypotheses, which constitute the proper empirical subject matter of science. These are mechanical hypotheses; they refer to configurations of matter in motion. As Descartes explains in Part III, 46, of the *Principia*, since it is a priori possible that there are countless configurations of matter in motion that can underlie the various natural phenomena, 'unaided reason' is not able to figure out the right configuration of matter in motion. Mechanical hypotheses are necessary but experience should be appealed to, in order to pick out the correct one: '[W]e are now at liberty to assume anything we please [about the mechanical configuration], provided that everything we shall deduce from it is {entirely} in conformity with experience' (III, 46; 1982, 106).

These mechanical hypotheses aim to capture the putative causes of the phenomena under investigation (III, 47). Hence, they are explanatory of the phenomena. Causal explanation – that is, mechanical explanation – proceeds via *decomposition*. It is a commitment of the mechanical philosophy that the behaviour of observable bodies should be accounted for on the basis of the interactions among their constituent parts and particles, hence, on the basis of unobservable entities. Descartes states (IV, 201; 1982, 283) that sensible bodies are composed of insensible particles. But to get to know these particles and their properties a *bridge principle* is necessary, that is, a principle that connects the micro-constituents with the macro-bodies. According to this principle, the properties of the minute particles should be modelled on the properties of macro-bodies. Here is how Descartes put it:

> Nor do I think that anyone who is using his reason will be prepared to deny that it is far better to judge of things which occur in tiny bodies (which escape our senses solely because of their smallness) on the model of those which our senses perceive occurring in large bodies, than it is to devise I know not what new things, having no similarity with those things which are observed, in order to give an account of those things [in tiny bodies]. {E. g., prime matter, substantial forms, and all that great array of qualities which many are accustomed to assuming; each of which is more difficult to know than the things men claim to explain by their means}. (IV, 201; 1982, 284)

In this passage Descartes does two things. On the one hand, he advances a *continuity thesis*: it is simpler and consonant with what our senses reveal to us to assume that the properties of micro-objects are the same as the properties of macro-objects. This continuity thesis is primarily *methodological*. It licenses certain kinds of explanations: those that endow matter in general, and hence the unobservable parts of matter, with the properties of the perceived bits of matter. It therefore licenses as explanatory certain kinds of unobservable configurations of matter, namely, those that resemble perceived configurations of matter. On the other hand, however, Descartes circumscribes mechanical explanation by noting *what it excludes*, that is, by specifying what does not count as a proper scientific explanation. He's explicit that the Aristotelian-scholastic metaphysics of substantial forms and powerful qualities is precisely what is abandoned as explanatory by the mechanical philosophy.[2]

All this was followed in the investigation of the mechanism of gravity and the (in)famous vortex hypothesis according to which the planets are carried by vortices around the sun. A vortex is a specific configuration of matter in motion – matter revolving around a centre. The underlying mechanism of the planetary system then is a system of vortices:

> [T]he matter of the heaven, in which the Planets are situated, unceasingly revolves, like a vortex having the Sun as its center, and . . . those of its parts

---

[2] Descartes accepts that scientific explanation does not require the truth of the claims about the micro-constituents of things (IV, 204; 1982, 286). In the next paragraph, however, he argues that his explanations have 'moral certainty' (IV, 205; 1982, 286–7).

which are close to the Sun move more quickly than those further away; and ... all the Planets (among which we {shall from now on} include the Earth) always remain suspended among the same parts of this heavenly matter. (III, 30; 1982, 96)

The very idea of this kind of configuration is suggested by experience, and by means of the bridge principle it is transferred to the subtle matter of the heavens. Hence, invisibility doesn't matter. The bridge principle transfers the explanatory mechanism from visible bodies to invisible bodies. More specifically, the specific continuity thesis used is the motion of 'some straws {or other light bodies} ... floating in the eddy of a river where the water doubles back on itself and forms a vortex as it swirls'. In this kind of motion we can see that the vortex carries the straws 'along and makes them move in circles with it'. We also see that 'some of these straws rotate about their own centers, and that those which are closer to the center of the vortex which contains them complete their circle more rapidly than those which are further away from it'. More importantly for the explanation of gravity, we see that 'although these whirlpools always attempt a circular motion, they practically never describe perfect circles, but sometimes become too great in width or in length.' Given the continuity thesis, we can transfer this mechanical model to the motion of the planets and 'imagine that all the same things happen to the Planets; and this is all we need to explain all their remaining phenomena' (III, 30; 1982, 96). Notably, the continuity thesis offers a heuristic for discovering plausible mechanical explanations.

Huygens (1690/1997) came to doubt the vortex theory, 'which formerly appeared very likely' to him (p. 32). He didn't thereby abandon the key tenet of mechanical philosophy. For Huygens too the causal explanation of a natural phenomenon had to be mechanical. He said, referring to Descartes: 'Mr Descartes has recognized, better than those that preceded him, that nothing will be ever understood in physics except what can be made to depend on principles that do not exceed the reach of our spirit, such as those that depend on bodies, deprived of qualities, and their motions' (pp. 1–2).

Huygens posited a fluid matter that consists of very small parts in rapid motion in all directions and which fills the spherical space that includes all heavenly bodies. Since there is no empty space, this fluid matter is more easily moved in circular motion around the centre, but not all parts of it move in the same direction. As Huygens put it 'it is not difficult now to explain how gravity is produced by this motion' (p. 16). When the parts of the fluid matter encounter some bigger bodies, like the planets, 'these bodies [the planets] will necessarily be pushed towards the center of motion, since they do not follow the rapid motion of the aforementioned matter'. And he added: 'This then is in all likelihood what the gravity of bodies truly consists of: we can say that this is the endeavor that causes the fluid matter, which turns circularly around the center of the Earth in all directions, to move away from the center and to push in its place bodies that do not follow this motion.' In fact, Huygens devised an experiment with bits of beeswax to show how this movement towards the centre can take place.

Newton of course challenged all this, along the lines that the very idea of causal explanation should be *mechanical*. But before we take a look at his reasons and their importance for the very idea of mechanical explanation, we should not fail to see the broader metaphysical grounding of the mechanical project. For, as we noted, in the seventeenth century mechanism offered the metaphysical foundation of science.

1.3.2   Mechanical versus Non-Mechanical Explanation

The contours of this endeavour are well known. Matter and motion are the 'ultimate constituents' of nature, or, as Boyle (1991, 20) put it, the 'two grand and most catholic principles of bodies'. Hence, all there is in nature (but clearly not the Cartesian minds) is determined (caused) by the mechanical affections of bodies and the mechanical laws. Here is Boyle again: '[T]he universe being once framed by God, and the laws of motion being settled and all upheld by his incessant concourse and general providence, the phenomena of the world thus constituted are physically produced by the mechanical affections of the parts of matter, and what they operate upon one another according to mechanical laws' (1991, 139).

The Boylean conception, pretty much like the Cartesian, took it that the new mechanical approach acquired content by excluding the then dominant account of explanation in terms of 'real qualities': the scholastics 'attribute to them a nature distinct from the modification of the matter they belong to, and in some cases separable from all matter whatsoever' (pp. 15–16). Explanation based on real qualities, which are distinct (and separable) from matter, is not a genuine explanation. They are posited without 'searching into the nature of particular qualities and their effects' (p. 16). They offer sui generis explanations: why does snow dazzle the eyes? Because of 'a quality of whiteness that is in it, which makes all very white bodies produce the same effect' (p. 16). But what is whiteness? No further story about its nature is offered, but just that it's a 'real entity' inhering in the substance. Why do white objects produce this effect rather than that? Because it is in their nature to act thus.

Descartes made this point too when, in his *Le Monde*, he challenged the scholastic rivals to explain how fire burns wood, if not by the incessant and rapid motion of its minute parts. In his characteristic upfrontness, Descartes contrasted two ways to explain how fire burns wood. The first is the Aristotelian way, according to which 'the 'form' of fire, the 'quality' of heat and the 'action' of burning' are 'very different things in the wood' (Descartes 2004, 6). The other is his own mechanistic way: when the fire burns wood, 'it moves the small parts of the wood, separating them from one another, thereby transforming the finer parts into fire, air, and smoke, and leaving the larger parts as ashes' (2004, 6).

This causal explanation, based as it is on matter in motion, is preferable precisely because it is explanatory of the burning; in contrast, the Aristotelian is not, precisely because it does not make clear the mechanism by which the fire consumes the wood: '[Y]ou can posit "fire" and "heat" in the wood and make it burn as much as you please: but if you do not suppose in addition that some of its parts move or are detached from their neighbours then I cannot imagine that it would undergo any alteration or change' (p. 6). To the then dominant account of real qualities, the new mechanical metaphysics juxtaposed a different view of qualities. For something to be a quality it should be determined by the mechanical affections of matter, that is, by 'virtue of the motion, size, figure, and contrivance, of their own parts' (Boyle 1991, 17). Hence, there can be no change in qualities unless there is a change in mechanical affections. Though 'catholic or universal matter' is common to all bodies (being, as Boyle [p. 18] put it, 'a substance extended, divisible, and impenetrable'), it is diversified by motion, which is regulated by laws.

The key point then is that the mechanical account of nature is both a metaphysical grounding of science *and* a (the) way to do science: offering mechanical explanations of the phenomena. It covers everything, from the very small to the very large. Here is Boyle again: 'For both the mechanical affections of matter are to be found, and the laws of motion take place, not only in the great masses and the middle-sized lumps, but in the smallest fragments of matter; and a lesser portion of it, being as well a body as a greater, must, as necessarily as it, have its determinate bulk and figure' (p. 143).

The metaphysical grounding of mechanical explanation renders it a distinct kind of explanation, which separates it sharply from rival accounts. Concomitantly, it becomes very clear what counts as a non-mechanical alternative. An explanation couched in terms of 'nature, substantial forms, real qualities, and the like' is 'unmechanical' (p. 142). But a sui generis chemical account of nature is unmechanical too. As Boyle put it:

> [T]hough chemical explications be sometimes the most obvious and ready, yet they are not the most fundamental and satisfactory: for the chemical ingredient itself, whether sulphur or any other, must owe its nature and other qualities to the union of insensible particles in a convenient size, shape, motion or rest, and contexture, all which are but mechanical affections of convening corpuscles. (p. 147)

The opposition to both of these non-mechanical accounts is weaved around a certain metaphysical account of the world as fundamentally mechanical and a reductive-decompositional account of scientific explanation itself.

1.3.3   Boyle on Mechanical Explanation

Boyle's discussion of the nature of mechanical explanation in his 'About the Excellency and Grounds of the Mechanical Hypothesis' deserves further analysis, as it is particularly relevant for our purposes in this book. In that essay, Boyle contrasts mechanical with other kinds of explanations and points out that only the former provide information about how exactly a result is produced. What is particularly interesting for us is that Boyle focuses on what can be broadly described as 'medical' examples. Consider the following passage:

> They that, to solve the phenomena of nature, have recourse to agents which, though they involve no self-repugnancy in their very notions, as many of the judicious think substantial forms and real qualities to do, yet are such that we conceive not how they operate to bring effects to pass – these, I say, when they tell us of such indeterminate agents as the soul of the world, the universal spirit, the plastic power, and the like, though they may in certain cases tell us some things, yet they tell us nothing that will satisfy the curiosity of an inquisitive person, who seeks not so much to know what is the general agent that produces a phenomenon, as *by what means, and after what manner, the phenomenon is produced.* (p. 144, emphasis added)

Here, Boyle points out that, in giving an explanation, it is not enough to state what the cause is; what is more important is to state how exactly a cause operates to bring about the effect. Failure to do this, Boyle thinks, is the main problem with explanations that merely state a causal agent without providing further information as to the manner that this agent acts. Boyle goes on to give an example of such an unsatisfactory medical explanation:

> The famous Sennertus and some other learned physicians tell us of diseases which proceed from incantation: but sure it is but a very slight account that a sober physician, that comes to visit a patient reported to be bewitched, receives of the strange symptoms he meets with and would have an account of, if he be coldly answered that it is a witch or the devil that produces them. (p. 144)

Similarly,

> it would be but little satisfaction to one that desires to understand the causes of what occurs to observation in a watch, and how it comes to point at and strike the hours, to be told that it was such a watchmaker that so contrived it; or to him that would know the true cause of an echo to be answered that it is a man, a vault, or a wood, that makes it. (p. 144)

The point that Boyle makes in these passages is that, quite apart from the accusation that notions such as substantial forms and real qualities are obscure, such explanations as well as others (e.g., in terms of plastic powers, which Boyle thinks are not as bad as the ones offered by the scholastics) do not fulfil what he takes as *a general adequacy condition* that an explanation has to satisfy, that is, to provide *information* as to how exactly a cause acts. Consider, for example, the following causal claim: administering of the poison led to the death of the person. We can interpret Boyle as saying that, qua explanation of death, such an explanation is incomplete; what is missing is how exactly administering of the poison led to death. Stating that the poison possessed the power to bring about death is tantamount to saying that it in fact produced death, that is, that it was the cause of death. What we need in addition to this, however, is *the way* it did so. A way to satisfy this demand is to provide information about the changes that the poison produced within the organism and explain how they eventually led to death. Here is how Boyle puts it:

> I consider that the chief thing that inquisitive naturalists should look after in the explicating of difficult phenomena is not so much what the agent is or does, as what changes are made in the patient to bring it to exhibit the phenomena that are proposed, and *by what means, and after what manner, those changes are effected*: so that, the Mechanical philosopher being satisfied that one part of matter can act upon another but by virtue of local motion or the effects and consequences of local motion, he considers that as, if the proposed agent be not intelligible and physical, it can never physically explain the phenomena, so, *if it be intelligible and physical, it will be*

*reducible to matter and some or other of those only catholic affections of matter
already often mentioned.* (p. 145, emphasis added)

According to Boyle, then, what an 'inquisitive person' should do in
offering an explanation of how an outcome such as death by poison comes
about is to describe the series of changes that led from the event that
counts as the cause to the resulting outcome. Moreover, such an explanation should be given in mechanical terms. The reason is that only by
means of local motion can we understand how a cause (i.e., a part of
matter) can operate on something (i.e., another part of matter). So, non-
physical causes cannot explain (since they do not act by means of local
motion), and for physical causes to be explanatory, they have to be reduced
to matter and to the 'catholic affections' of matter, including motion. The
requirement that an adequate explanation has to state the means by which
the cause acts is here supplemented by the further requirement that the
account of how exactly it acts has to be given in mechanical terms. We can
view the resulting account of explanation as a combination of a methodological thesis (i.e., that an explanation should state how exactly the cause
acts) with an ontological one (i.e., that it should be given in mechanical
terms). Boyle goes on:

> whatever be the physical agent, whether it be inanimate or living, purely
> corporeal or united to an intellectual substance, the above-mentioned
> changes, that are wrought in the body that is made to exhibit the phenom-
> ena, may be effected by the same or the like means, or after the same or the
> like manner ... And if an angel himself should work a real change in the
> nature of a body, it is scarce conceivable to us men how he could do it
> *without the assistance of local motion,* since, if nothing were displaced or
> otherwise moved than before (the like happening also to all external bodies
> to which it related), it is hardly conceivable how it should be in itself other
> than just what it was before. (p. 146, emphasis added)

Boyle argues here that since real change requires local notion, a mechanical
explanation can always be given no matter what the exact nature of the
agent is. Even if the agent is immaterial, to produce a certain outcome is to
produce a series of changes in local motion. This leads, finally, to the
following point:

> From the foregoing discourse it may (probably at least) result that if, besides
> rational souls, there are any immaterial substances (such as the heavenly
> intelligences and the substantial forms of the Aristotelians) that regularly are
> to be numbered among natural agents, their way of working being
> unknown to us, they can but help to constitute and effect things, but will
> very little help us to conceive how things are effected: so that, *by whatever*

principles natural things be constituted, it is by the Mechanical principles that their phenomena must be clearly explicated. (p. 150, emphasis added)

Boyle here points out that even if substantial forms were to be accepted, they (as well as other immaterial substances) are useless if our aim is to understand how the phenomena are brought about. The thought here is that, in order to explain phenomena, we have to explain how they come about; but we can do this only by stating how the various causal agents produce changes in the properties of parts of matter by means of local motion; we cannot conceive how substantial forms can result in local motion; so, substantial forms are useless in offering explanations of how phenomena are produced.

In sum, Boyle's main thought is that a satisfactory causal explanation has to explain the way a cause acts in bringing about a certain phenomenon. To do this, a causal explanation has to be mechanical, where a mechanical explanation explains how the effect is brought about in terms of changes in the mechanical properties of parts of matter, including local motion. As we will explain in Section 1.4, in giving our account of mechanism and mechanistic explanation, we will be in agreement with Boyle's insights. But whereas we will keep his methodological thesis (i.e., that in giving a mechanistic explanation in science one has to explain how exactly a cause acts by describing the sequence of causal steps leading from the cause to the effect), we will reject the ontological thesis (i.e., that there exists a privileged description of this causal sequence, either in physico-chemical terms or in terms of one's favourite metaphysics of causal processes).

1.3.4 Newton against Mechanism

When Newton offered a non-mechanical account of gravity, he primarily challenged the idea that legitimate scientific explanation ought to be mechanistic, at least in the narrow sense of taking all action to be by *contact*. There is a sense in which Newton prioritised explanation by unification under laws and not by mechanisms. This is seen in the *Preface* to the second (1713) edition of the *Principia*, authored by Roger Cotes under the supervision of Newton. In this preface, Cotes presents Newton's method as a middle way (*via media*) between Aristotelianism and Mechanism. To be sure, the mechanical explanations offered by the Cartesians were an improvement over the Scholastic explanations because they relied on demonstrations on the basis of laws. Still, taking 'the foundation of their speculations from hypotheses', the mechanists are 'merely putting together a romance [i.e., fiction], elegant perhaps and charming, but nevertheless a romance' (Newton 2004, 43).

Thus put, the point sounds epistemic; it concerns the increased risk involved in hypothesising a mechanism which is supposed to underpin, and hence to causally explain, a certain phenomenon. Cotes adds:

> But when they [the mechanists] take the liberty of imagining that the unknown shapes and sizes of the particles are whatever they please, and of assuming their uncertain positions and motions, and even further of feigning certain occult fluids that permeate the pores of bodies very freely, since they are endowed with an omnipotent subtlety and are acted on by occult motions: when they do this, they are drifting off into dreams, ignoring the true constitution of things, which is obviously to be sought in vain from false conjectures, when it can scarcely be found out even by the most certain observations. (p. 43)

Still, it's fair to say that Newton's *via media* was based on a different understanding of scientific explanation: it should certainly look for causes – hence, scientific explanation should be causal – but the sought-after causes need not act by the principles of Mechanism. Newton's way, Cotes says, is to 'hold that the causes of all things are to be derived from the simplest possible principles', but unlike the mechanists' way, it 'assume(s) nothing as a principle that has not yet been thoroughly proved from phenomena'. The 'explication of the system of the world most successfully deduced from the theory of gravity' is the 'most illustrious' example of Newton's way (p. 32).

Newton emphatically denied feigning any hypotheses about the cause of gravity. For him, 'it is enough that gravity really exists and acts according to the laws that we have set forth and is sufficient to explain all the motions of the heavenly bodies and of our sea' (p. 92). Gravity according to Newton is a non-mechanical force since it 'acts not in proportion to the quantity of the *surfaces* of the particles on which it acts (as mechanical causes are wont to do) but in proportion to the quantity of *solid* matter, and whose action is extended everywhere to immense distances, always decreasing as the squares of the distances' (p. 92). He added that the very motion of the comets makes it plausible to think that the regular elliptical motion of the planets (as well as of their satellites) cannot 'have their origin in mechanical causes' (p. 90).

In his already mentioned *Discourse on the Cause of Gravity* (1690), Huygens expressed his dissatisfaction with Newton's failure to offer a *mechanical* explanation of the cause of gravitational attraction. Favouring his own explanation of gravity in terms of the centrifugal force of the subtle and rapidly moving matter that fills the space around the Earth and the other planets, Huygens noted that Newton's theory supposes that gravity is 'an inherent quality of corporeal matter'. 'But', he immediately added, such a hypothesis 'would distance us a great deal from mathematical or mechanical principles' (1690/1997, 35).

Yet Huygens had no difficulty in granting that Newton's law of gravity was essentially correct when it comes to accounting for the planetary system. As he put it:

> I have nothing against *Vis Centripeta*, as Mr. Newton calls it, which causes the planets to weigh (or gravitate) toward the Sun, and the Moon toward the Earth, but here I remain in agreement without difficulty because not only do we know through experience that there is such a manner of attraction or impulse in nature, but also that it is explained by the laws of motion, as we have seen in what I wrote above on gravity. (p. 31)

Explaining the fact that gravity depends on the masses and diminishes with distance 'in inverse proportion to the squares of the distances from the centre' (p. 37) was, for Huygens, a clear achievement of Newton's theory despite the fact that the mechanical cause of gravity remained unidentified.

Commitment to mechanical explanation was honoured by Gottfried Wilhelm Leibniz too. In a piece titled 'Against Barbaric Physics: Toward a Philosophy of What There Actually Is and against the Revival of the Qualities of the Scholastics and Chimerical Intelligences' (written between 1710 and 1716), he defended the mechanical view by arguing that corporeal forces should be grounded mechanically when it comes to their application to the natural world. Leibniz was very clear that though he allowed 'magnetic, elastic, and other sorts of forces', they are permissible 'only insofar as we understand that they are not primitive or incapable of being explained, but arise from motions and shapes' (Leibniz 1989, 313). So, forces are necessary, but a condition for their applicability to the natural world is that they are seen as 'arising from motions and shapes'. What he took it to be 'barbarism in physics' was to posit sui generis, that is non-mechanically grounded, 'attractive and repulsive' forces that act at a distance (pp. 314–15). Newton's gravity was supposed to be such a barbaric force!

In a letter he sent to Nicolaas Hartsoeker (Hanover, 10 February 1711), Leibniz makes it clear that the proper scientific explanation should be mechanical. It is not enough for scientific explanation to identify the law by means of which a certain force acts; what is also required is the specification of the mechanism by means of which it acts. The mechanism is, clearly, on top of the law and given independently of it. Without the mechanism the power is 'an unreasonable occult quality'. He says:

> Thus the ancients and the moderns, who own that gravity is an occult quality, are in the right, if they mean by it that there is a certain mechanism unknown to them, whereby all bodies tend towards the center of the earth. But if they mean that the thing is performed without any mechanism by a simple primitive quality, or by a law of God, who produces that effect without using any intelligible means, it is an unreasonable occult quality, and so very occult, that it is impossible it should ever be clear, though an angel, or God himself, should undertake to explain it. (Newton 2004, 112)

Newton couldn't disagree more. In an unsent letter written circa May 1712 to the editor of the *Memoirs of Literature*, Newton referred explicitly to Leibniz's letter to Hartsoeker and stressed that it is not necessary for the introduction of a power – such as gravity – to specify anything other than the law it obeys; no extra requirements should be imposed, and in particular no requirement for a mechanical grounding:

> And therefore if any man should say that bodies attract one another by a power whose cause is unknown to us, or by a power seated in the frame of nature by the will of God, or by a power seated in a substance in which bodies move and float without resistance and which has therefore no vis inertiae but acts by other laws than those that are mechanical: I know not why he should be said to introduce miracles and occult qualities and fictions into the world. For Mr. Leibniz himself will scarce say that thinking is mechanical as it must be if to explain it otherwise be to make a miracle, an occult quality, and a fiction. (Newton 2004, 116)

Note well Newton's point. The fact that an explanation does not conform to a certain mechanical framework does not make it fictitious, occult or miraculous. Non-mechanical explanations are legitimate insofar as they identify the law that covers or governs a certain phenomenon. Hence, Newton promotes a methodological shift: causal explanation without mechanisms but subject to laws.

Causal explanation then need not be mechanical to be legitimate and adequate. This is Newton's key thought. In breaking with a tradition which brought under the same roof a certain metaphysical conception of the world and a certain view of scientific explanatory practice, Newton distinguished the two and laid emphasis on the explanatory practice itself, thereby freeing it from a certain metaphysical grounding.

Though this is not the end of the story of Old Mechanism (more will be said in the next chapter), Newton's key thought, we shall argue, is of relevance in the current debates over New Mechanism, to which we shall now turn our attention.

1.4 New Mechanism: From Practice to Metaphysics

It is useful to differentiate between two ways to conceptualise mechanisms in the post-1970 literature. First, mechanism has been used as a primarily metaphysical concept, mostly aiming to illuminate the metaphysics of causation. Second, mechanism has been taken to be a concept used in science, and philosophical accounts of mechanism have aimed to elucidate this concept.

To be sure, some philosophical approaches to mechanism, most notably Glennan's (1996), blend these two conceptions (the metaphysical one and the concept-in-use). However, it's fair to say that there are two quite distinct points of origin of the recent philosophical accounts of mechanism: the first starts from metaphysics (as was the case for Descartes and other old mechanists), the second from scientific practice. Using this distinction between mechanism as a primarily metaphysical concept and as a concept-in-use in science, we can differentiate between two kinds of approaches to the metaphysics of mechanisms.

On the first approach, the aim is to show what the connection is between mechanism qua a metaphysical category and other central metaphysical concepts, notably, causation. In the context of the metaphysics of causation, 'mechanistic' accounts are theories about the link between cause and effect. Such theories are meant to be anti-Humean in that they view causation as a productive relation; that is, the cause somehow brings about or produces the effect. The aim of the mechanistic view of causation is to illuminate the productive relation between the cause and the effect by positing a mechanism that connects them and by explicating 'mechanism' in a suitable way such that causal sequences are differentiated from non-causal ones. The central thought, then, is that A causes B if and only if there is a mechanism connecting A and B.

Two kinds of views have become prominent: those that characterise the mechanism that links cause and effect in terms of the persistence, transference or possession of a conserved quantity (Mackie 1974; Salmon 1997; Dowe 2000) and those that connect a mechanistic account to causal production with a power-based one (see Harré 1970 for an early such view). Despite their differences, these views share in common the claim that mechanisms are the ontological tie that constitutes Hume's 'secret connexion'. We call such mechanisms *mechanisms-of*. Mechanisms-of are ontological items that underlie or constitute certain kind of processes, that is, those that can be deemed causal. We will deal with these accounts in more detail in Chapter 5 (see also Psillos 2002).

On the second approach, working out a metaphysics of mechanisms is not the starting point but rather the end point of inquiry. Starting with mechanism as a concept-in-use in science, one tries first to give a general characterisation of this concept and then to derive metaphysical conclusions, that is, conclusions about the (mechanistic) structure of the world. This kind of bottom-up inquiry has yielded several well-known general accounts of mechanisms as well as theses about the ontic signature of a mechanistic world.

1.4.1   The Metaphysics of New Mechanism

Here are three well-known general characterisations of a mechanism in recent mechanistic literature:

> Mechanisms are entities and activities organized such that they are productive of regular changes from start or set-up to finish or termination conditions. (Machamer et al. 2000, 3)

> A mechanism for a behavior is a complex system that produces that behavior by the interaction of a number of parts, where the interactions between parts can be characterized by direct, invariant, change-relating generalizations. (Glennan 2002, S344)

> A mechanism is a structure performing a function in virtue of its component parts, component operations, and their organization. The orchestrated functioning of the mechanism is responsible for one or more phenomena. (Bechtel & Abrahamsen 2005, 423)

The focus on mechanism as a concept-in-use is common to all three accounts; none of the three accounts can be viewed as falling under the rubric of mechanistic theories of causation. And yet, all these and similar accounts yield specific metaphysical commitments about what kind of things in the world mechanisms are. All these accounts are committed to the thesis that a general characterisation of mechanism must itself be cashed out in metaphysical terms.[3] Hence, talk of mechanisms in science is taken to have quite direct consequences about the kind of ontology

---

[3] As Bechtel and Abrahamsen are more interested in offering an account of explanation rather than saying what mechanisms are as things in the world, their characterisation of mechanism can be read in a metaphysically deflationary way.

presupposed by such talk. In order to substantiate this point, let us look at the three accounts mentioned earlier in some more detail.

Peter Machamer, Lindley Darden and Carl Craver's (henceforth MDC) account is perhaps the most ontologically inflated, as it is explicitly committed to both entities *and* activities as distinct and separate ontological categories. It is thus committed to a particular view about the metaphysics of causation: causation within mechanisms is to be characterised in terms of *production*, where the productive relation is captured by the various different kinds of activities identified by science.

Glennan's case is interesting, since in his 2002 article he refrains from taking mechanisms to entail a productive account of causation. Instead, within-mechanism interactions are characterised in terms of invariant, change-relating generalisations. As we will see below, however, Glennan has connected his account of mechanisms with a power-based understanding of causation. Hence, he is committed to causal powers as parts of the building blocks of mechanisms.

Last, Bechtel and Abrahamsen's account does not include a specific characterisation of what mechanistic causation amounts to at all. Here, however, as in the other two accounts, we have a series of general terms, the meaning of which needs to be unpacked. So, MDC include in their accounts 'entities' and 'organisation'; Glennan in his early formulations includes 'complex system' and 'parts'; and Bechtel and Abrahamsen talk about 'structure', 'function', 'parts' and 'organisation'.

All these accounts suggest the further need to explain what this 'new mechanical ontology' of entities, activities, organisation of parts into wholes and so on amounts to; what, in general terms, the constituents of mechanisms are and what their relations are to more traditional metaphysical categories, such as things, properties, powers and processes.

Notably, there has been a tendency recently to offer a more minimal general characterisation of a mechanism. For example, according to Illari and Williamson:

> A mechanism for a phenomenon consists of entities and activities organized in such a way that they are responsible for the phenomenon. (2012, 120)

Glennan's recent version is almost identical:

> A mechanism for a phenomenon consists of entities (or parts) whose activities and interactions are organised so as to be responsible for the phenomenon. (2017, 17)

Glennan calls this account *Minimal Mechanism*. The key motivation here is for a general characterisation of mechanism broad enough to capture examples of mechanisms in different fields, from physics to the social sciences. But even in this minimal mode, mechanisms, according to Glennan, 'constitute the causal structure of the world' (2017, 18).

This minimal account of mechanism might appear to fit the bill of capturing a concept-in-use in science. On closer inspection, however, it is committed to a rather rich metaphysical account of mechanism: the minimal account is not more minimal than the metaphysically inflated accounts noted above. The reason is that both of the foregoing minimal accounts still invite questions about the ontic status of mechanisms. For example, how exactly do entities and activities differ? What is the relation between activities and interactions? How should organisation be understood? Glennan (2017, 13) explicitly talks about a 'new mechanical ontology' as the upshot of the minimal account. The 'minimal mechanism', he adds, 'is an ontological characterization of what mechanisms are as things in the world' (p. 19).

New Mechanism, then, aims to provide a new ontology of mechanisms. We can identify three commonly accepted key theses concerning mechanical ontology:

(1)   The world consists of mechanisms.

Thesis 1 is a typical view among mechanists: mechanisms are taken to be *things in the world*, with (more or less) objective boundaries.[4] Ours is a mechanistic world. As Glennan puts it, '[t]hat is just how we have found the world to be' (2017, 240).

(2)   A mechanism consists of objects of diverse kinds and sizes structured in such a way that, in virtue of their properties and capacities, they engage in a variety of different kinds of activities and interactions such that a certain phenomenon P is brought about.

Thesis 2 (or something very similar) can be taken as the common core of the general characterisations of mechanism as a concept-in-use given by new mechanists. It identifies the components of a mechanism and the relations among them. As mechanisms are things in the world (thesis 1), their components are also particular things in the world. Besides, these parts engage in activities by being 'active, at least potentially' (Glennan 2017, 21). Activity is understood as a manifestation of the powers things have. Glennan is quite explicit that 'Activities manifest the powers

---

[4] The issue of boundaries is far from settled in the literature; we will come back to this issue in Chapter 8.

(capacities) of the entities involved in the Activity' (p. 31). Positing powers is supposed to explain why 'activities are powerful'; being powerful, activities are what 'an entity does, not merely something that happens to an entity' (p. 32). But activities are not enough. Interactions are needed too because 'there is no production without interaction' (p. 22). 'The fundamental point of ontological agreement among the New Mechanists', as Glennan (2017, 21 n. 6) puts it, is that entities cannot exist without activities or activities without entities. It's not hard to see that the minimal account of mechanism is taken to imply or suggest a rather substantive metaphysical conception of mechanism, which, until further notice at least, is broadly neo-Aristotelian.

(3)   The main way to explain a certain phenomenon P is to offer the mechanism that produces it.

Thesis 3 connects the previous theses with a claim about explanation (and more specifically, causal explanation): since in a mechanistic world phenomena are produced by mechanisms, the main task of scientific explanation is to identify the mechanism that produces a certain phenomenon, that is (by thesis 2), to identify the organised entities and activities that produce the phenomenon.

Despite their differences, there are important similarities between Old and New Mechanism (which justify viewing both positions as mechanistic). On the one hand, as we saw, new mechanists differ from their seventeenth-century predecessors in that they do not start their analysis with a metaphysical concept of mechanism; rather, they aim at giving a general characterisation of mechanism as a central concept of scientific practice. This characterisation is non-reductive in that it is not committed to the view that mechanisms are configurations of matter in motion subject to laws (and contact action). But, on the other hand, they are committed to mechanisms being configurations of powerful entities engaged in activities and interactions. As Glennan puts it: 'Mechanisms are particular and compound, made up of parts (entities) whose activities and interactions are located in particular regions of space and time' (2017, 57). Hence, New Mechanism is similar to seventeenth-century Mechanism, in that it is committed to a mechanical *ontology*. This ontology (theses 1 and 2 above), while not a global metaphysics in the sense of the seventeenth century, is still a thesis about the ontic signature of the world. Here is Glennan again: 'New Mechanist ontology is an ontology of compound systems. It suggests that the properties and activities of things must be explained by reference to the activities and organization of their parts' (p. 57). Instead of resulting in a 'flat' ontology where everything there is consists in matter in motion, this new mechanical metaphysics ends up with a hierarchy of particular things – mechanisms – which may contain a diverse set of entities and activities, rather than the limited set endorsed by the corpuscularians, and whose productivity is grounded in causal powers, rather than in a few fundamental laws of motion.

But we can ask: Are these ontological commitments really necessary in order to understand scientific practice? Are they licensed by the practice of science? Remember here that the primary aim of new mechanists is to give a general characterisation of mechanism as a concept-in-use. So, ideally, the general account of mechanism should capture as far as possible the extension of a concept-in-use in the various sciences. The minimal account of mechanism discussed so far, though broad enough to play this role, inflates the concept-in-use by making it amenable to a certain metaphysical description of its basic components.

Note that our claim is not that the metaphysical questions are not philosophically interesting questions to ask; they are, especially if we are interested in giving an account of the ontological structure of reality. Moreover, such a kind of project has to be informed by what science has to say about the world. If, however, our aim is to understand how a specific concept – *mechanism* – is used in scientific practice, these questions seem, at least prima facie, irrelevant, especially if a general characterisation of mechanism is possible that does not include such things.

1.4.2   *Mechanism in Scientific Practice*

A metaphysically deflationary view of mechanism as a concept-in-use that is broad enough to capture all examples of mechanisms that we find in science seems indeed possible. This is skinnier than those accounts of mechanism offered by Illari and Williamson and by Glennan. We nonetheless claim that this skinny account is enough to capture the concept-in-use. It will be the main aim of Chapter 4 to present this account in detail, and the subsequent chapters will further illuminate various features of the view. Here we will just introduce the basic idea behind our account. This skinny or, as we will prefer to describe it, *deflationary* account of mechanism is achieved by dropping the reference to activities and interactions and by understanding mechanism as the causal pathway of a certain phenomenon, described in the language of theory. According to this account that we call *Causal Mechanism* (CM), a mechanism in science just is a causal pathway described in theoretical language:[5]

(CM) A mechanism is a theoretically described causal pathway.

The central idea behind CM stems from the Boylean insight introduced in Section 1.3.3: when scientists talk about a 'mechanism', what they try to capture is the way (i.e., the causal pathway) a certain result is produced. Say, for example, that a pathologist tries to find out how a certain disease state is brought about. They will look for a specific mechanism, that is, a causal pathway that involves various causal links between, for example, a virus infection and changes in properties of the organism that ultimately lead to the disease state. In pathology such causal pathways are referred to as the 'pathogenesis' of a disease, and when pathologists talk about the 'mechanisms' of a disease, it is such pathways that they have in mind (see Lakhani et al. 2009).

According to CM, then, mechanisms and causation are closely related: when two events are causally connected, there is a mechanism (i.e., a causal pathway) that connects them and accounts for the specific way that the cause brings about the effect. Scientists succeed in identifying a mechanism, if they succeed in describing the relevant causal pathway in terms of the theoretical language of the particular scientific field. An especially clear example of the identification of a new mechanism is the case of the mechanism of cell death known as apoptosis; we will examine this case in detail in the next chapter.

The view of mechanisms as causal pathways differentiates CM from accounts that explicitly view mechanisms as complex systems (Glennan 1996), kinds of structures (Bechtel & Abrahamsen 2005) or more generally as organised entities of some sort; it doesn't differentiate it from more processual views, such as the MDC account. CM stresses that mechanisms are not systems, but causal processes. It is therefore closer to the older Salmon-Dowe view, as well as to the Boylean conception sketched above, than some more recent accounts.

There exists, however, a very important difference between Boyle's notion of mechanical explanation and CM. As we saw earlier, for Boyle

---

[5] We have first presented *Causal Mechanism* in Ioannidis and Psillos (2017), where we have called it *Truly Minimal Mechanism* to differentiate it from Glennan's *Minimal Mechanism*. But in this book we will use 'Causal Mechanism' to describe our account, as this indicates the close relationship between mechanism and causation, where causation is used to understand what a mechanism in science is, rather than vice versa (as in mechanistic theories of causation – we will examine these theories and the relations between mechanisms and causation in detail in Chapter 5).

and other mechanical philosophers of the seventeenth century, mechanical explanations had to be couched in very specific terms, that is, in terms of the changes produced by parts of matter to the 'mechanical affections' (including motion) of other parts of matter. Thus, the methodological claim of mechanical philosophers – that is, that to explain how the phenomena are produced one should identify the mechanisms that produce them – did incorporate a claim about the specific theoretical language that such explanations should be couched in. And the main justification of this latter claim was ontological: what really exists in the world is matter that behaves according to certain laws that govern its motion. Old Mechanism, then, combined a methodological claim about the preferred form of scientific explanation with an ontological claim, that is, a claim about how the world is constructed.

In contrast to this more restricted way to conceive of mechanistic explanation, according to CM there is no privileged theoretical language in terms of which the causal pathway that produces the phenomena has to be described. This is, in one sense, in agreement with the dominant views of what a mechanism is: as both MDC (Machamer et al. 2000) and Glennan (1996) stress in the papers that first offered general characterisations of the notion of mechanism, the contemporary concept is more general than its seventeenth-century counterpart, as the parts of a mechanism interact in various ways (e.g., by chemical interactions) and thus are not 'mechanisms' in the restricted seventeenth-century sense of the term.

In another sense, however, CM is different from current accounts, as it stresses that there is no privileged ontological description of a mechanism. So, while current accounts combine the methodological claim that science should discover the causal pathways that produce the phenomena with an ontological claim about the metaphysics of mechanisms, CM deflates the metaphysics and puts the methodological claim at the centre. We shall further examine this feature of CM by revisiting Newton's views. Before this, however, let us dispel a natural but pointed objection. Recall our main thesis that 'A mechanism is a theoretically described causal pathway.' Does this invite the interpretation that a mechanism does not exist before it is theoretically described? We, of course, do not believe this; our view is that mechanisms just are independently and objectively existing causal processes. The aim of the phrase 'theoretically described' is to highlight that the causal processes that constitute mechanisms are to be described in the theoretical terms of the relevant scientific domain and not in terms of general ontological categories.

1.5 Newton Revisited

What does Newton's critique of the Old Mechanism have to do with our understanding (and criticism) of New Mechanism? In a letter to Leibniz dated 16 October 1693, Newton challenged him to offer a mechanical explanation of 'gravity along with all its laws by the action of some subtle matter' and to show that 'the motion of planets and comets will not be disturbed by this matter'. If this were available, Newton said, he would be 'far from objecting'. But no such explanation was forthcoming and Newton was happy to reiterate his view that 'since all phenomena of the heavens and of the sea follow precisely, so far as I am aware, from nothing but gravity acting in accordance with the laws described by me; and since nature is very simple . . . all other causes are to be rejected' (Newton 2004, 108–9). Newton does not simply say that causal explanation might not be mechanical. His point is that causal explanation should be liberated from the tenets of (the narrowly understood) Old Mechanism. It would not be enough to offer a mechanical account of the cause of gravity; the laws that gravity obeys should be mechanically explicable, and, as Newton repeatedly stressed, this was not forthcoming. Though causal explanation matters, it doesn't matter if it is subject to various (old) mechanical constraints.

We noted already that the new mechanical conception of nature is far from the seventeenth-century conception that everything should be accounted for in terms of (configurations of) matter in motion. So it's far from us to tar New Mechanism with the same brush as Old Mechanism. For instance, the key ontology of the old mechanical picture was justified by and large a priori, whereas the key ontology of New Mechanism is grounded in scientific practice; in this case, it is practice that constrains metaphysics. Be that as it may, we are now going to argue that there exists a kind of Newtonian move against New Mechanism too.

What is clear from the present discussion is that, regardless of the main difference noted above, the new idea of mechanism is no less metaphysically loaded than the old one. Where the seventeenth-century mechanists looked for stable arrangements of matter in motion subject to laws, the twenty-first-century mechanists look for stable arrangements of powerful entities engaged in various activities and interactions. These mechanisms are supposed to be the building blocks of nature, and the scientific task is to unravel them. They underpin 'mechanistic explanations' which, as Glennan put it, show 'how the organized activities and interactions of some set of entities cause and constitute the phenomenon to be explained' (2017, 223). Mechanistic explanation 'always involves characterizing the activities and interactions of a mechanism's parts' (p. 223). Where the seventeenth-century mechanists saw 'action by contact' as a requisite for a proper mechanical explanation, new mechanists see powers and 'activities'.

Why is Newton's key thought relevant to the modern debates about mechanisms? The key thought, to repeat, was that causal explanation should identify causes and the laws that govern their action irrespective of whether or not these causes can be taken to satisfy further (mostly metaphysically driven) constraints. In other words, Newton showed that certain causal explanations of phenomena (in terms of non-mechanical forces) are both legitimate and complete insofar as they identify the right causes and are empirically grounded.

We take it that the point CM stresses, is, *mutatis mutandis*, analogous to Newton's. The point of CM is that causal explanation need not be mechanistic in the new mechanists' ontic sense and that being couched in the way new mechanists propose, causal explanation is subjected to constraints unwarranted by scientific practice. Insofar as mechanism is a concept-in-use in science, it may well be seen referring to the causal pathway of the phenomenon to be explained, couched in the language of theories. Preserving the spirit of Newton's key thought, we might say that causal explanation is legitimate even if we bracket the issue of 'what mechanisms or causes are as things in the world' (Glennan 2017, 12) or the issue of what activities are and how they are related to powers and the like. The issue then is not 'an ontological characterization of what mechanisms are as things in the world' (p. 19), but a methodological characterisation of them as causal pathways described in the language of theories.

To press the analogy a bit more, questions such as 'If entities, activities, and the mechanisms they constitute are compounds, of what are they compounded? Where does one entity or activity or mechanism end, and when does another begin? And on what account do we decide that a collection of interacting entities is to count as a whole mechanism?' (p. 29) are pretty much like the questions concerning the cause of the properties of gravity that Newton thought need not be asked and answered for a scientifically legitimate conception of causal explanation.

We do not want to claim that questions such as the above are not connected to scientific practice. After all, even the question of the cause of gravity that Newton refrained from answering was connected to scientific practice. The point, rather, we take from Newton is that answering these questions is not required for offering adequate causal explanations of the phenomena under study. Similarly, for CM, answering questions such as the above is not required in order to have legitimate mechanistic explanations. In other words, the properties of mechanism over and above those that are required by its methodological use need not be specified; nor is there an explanatory lacuna if they are not.

According to CM, the concept of mechanism as used in practice need not, and should not, be understood in a metaphysically inflated sense. Hence, new mechanists, in offering such metaphysically inflated accounts, need to show that such accounts are indeed indispensable for doing good mechanistic science. To conclude, as Newton remained agnostic about the underlying mechanism of gravity, so CM remains agnostic about the metaphysical ground of any particular causal pathway. As in the case of gravity, *it is enough that mechanisms qua causal pathways really exist and act as they do.*

CHAPTER 2

Extending Mechanism beyond the Two 'Most Catholic Principles of Bodies'

2.1 Preliminaries

In the previous chapter we offered a sketchy but hopefully intriguing account of the history of the mechanical world view. Now, similar stories have been told by the new mechanists themselves. A story about Newton occurs in Glennan (1992), and an account of the relationship between the austere Old Mechanism and richer New Mechanism shows up in a historical introduction to a journal issue by Craver and Darden (2005). Where does our story so far differ from theirs? Though all these accounts are very interesting, ours is better, we think, for three reasons. First, it is a lot more detailed than other accounts; second, and relatedly, we pay a lot more attention to historical accuracy compared with other accounts; third, and more importantly, we focus on an issue that is not discussed in detail in other accounts, that is, the reasons that led to the abandonment of Old Mechanism. Our analysis why Old Mechanism was abandoned leads to a fresh look at the differences and similarities between Old and New Mechanism and examines what the lessons are for current accounts.

Our view then is that history is important not so much because of the common elements between Old and New Mechanism, or as an 'origin story' of New Mechanism, but because of the lessons one can learn from a close examination, at both the metaphysical and the methodological levels, of the limitations and shortcomings of Old Mechanism when applied to the problem of explaining gravity. So, while for old mechanists such as Descartes and Leibniz scientific explanations of gravity (and scientific explanations in general) had to be couched in mechanical terms (i.e., in terms of matter in motion), with Newton this view was superseded. In fact, we have argued that according to Newton causal explanation should be liberated from the search for mechanisms. Would that be too strong a conclusion since, one might argue, for Newton the motion of the planets and the tides was mechanistic, even if there was not a mechanistic explanation (indeed any explanation at all) of the gravitational force itself?

Answering this potential objection requires delving into the rest of our story of the fate of the mechanical conception of the world. This will be part of the present chapter. Put in a nutshell, however, our reply is that the issue is partly terminological and partly substantive. It is substantive insofar as in the case of explanations that refer to forces acting at a distance (whether in the case of gravity or anything else), and for which no mechanical explanation can be given (i.e., their action cannot be explained in mechanical terms), we have causal explanations without mechanisms in the sense that natural philosophers like Leibniz or Huygens understood the notion. As Leibniz argues in his essay 'Against Barbaric Physics' (which we discussed briefly in Section 1.3.4), such explanations go against basic mechanistic tenets. So, it is clear to us that with Newton there is a fundamental change in what counts as a legitimate scientific explanation; consequently, if we want to designate Newton's explanations in terms of forces acting at a distance as 'mechanical', we have here a new and more liberal notion of mechanism. And that's why the issue is partly terminological. For as we shall show in the present chapter, there is a sense in which Newton did modify and expand the notion of mechanism prevalent in seventeenth century, which was further modified and extended by Kant and others after him.

2.2 Mechanical versus Quasi-Mechanical Mechanism

It will be helpful and accurate to distinguish between two concepts of mechanism – or, if you like, between two ideas of mechanism. We may call the first *mechanical* mechanism and the second *quasi-mechanical* mechanism. The first conception of mechanism is narrow: mechanisms are configurations of matter in motion subject to mechanical laws. It is this conception that has been associated with the rise and dominance of the mechanical conception of nature in the seventeenth century.

The second conception of mechanism is broader. A quasi-mechanical mechanism is *any* arrangement of parts into wholes in such a way that the behaviour of the whole depends on the properties of the parts and their mutual interactions. Harré (1972, 116) has called this kind of mechanism *generative* mechanism. The focus is not on the mechanical properties of the parts, nor on the mechanical principles that govern the behaviour of the parts and determine the behaviour of the whole. Instead, the focus is on the causal relations there are between the parts and the whole. Generative mechanisms are taken to be the bearers of causal connections. It is in virtue of them that the causes are supposed to produce the effects. There is a concomitant conception of mechanical explanation as a kind of decompositional explanation: an explanation of a whole in terms of its parts, their properties and their interactions. This second conception is, arguably, associated with Kant's idea of mechanism in his third critique.

2.2.1 Mechanical Mechanism

In the seventeenth century, as we explained in detail in the previous chapter, the mechanical conception of nature was taken to be a weapon against the Aristotelian view that each and every explanation was not complete unless some efficient *and* some final cause were cited. The emergent mechanical philosophy placed in centre-stage the new science of mechanics and left Aristotelian physics behind. Accordingly, the call for a mechanical explanation of phenomena has had definite content: all natural phenomena are *produced* by the mechanical interactions of the parts of matter according to mechanical laws.

The broad contours of the mechanical conception of nature were not under much dispute, at least among those who identified themselves as mechanical philosophers. The key ideas were that all natural phenomena are explicable mechanically in terms of matter in motion; that efficient causation should be understood, ultimately, in terms of *pushings* and *pullings*; and that final causation should be excised from nature.[1] Though definite, this conception was far from monolithic. As Marie Boas (1952) has explained in detail, there had been different and opposing conceptions as to the structure of matter (atomistic vs corpuscularian), the reality of the void (affirmation of the existence of empty space vs the plenum) and the primary qualities of matter (solely extension vs richer conceptions that include solidity, impenetrability and other properties). And yet the unifying idea was that all explanation is mechanical explanation and proceeds in terms of matter and motion.

With Newton, the content of the scientific conception of nature was altered and broadened. To the eyes of most of his contemporaries (notably to Leibniz and Huygens) Newton had just abandoned the principles of

---

[1] To be sure, most mechanical philosophers did find a role for final causation via God's design of the world, but crucially, this design was precisely that of a *mechanism*. More specifically, mechanical philosophers denied the presence in nature of immanent final causes such as Aristotelian forms. Indeed, an important characteristic of the mechanical conception of nature was its denial of *forms* as part of the acceptable ontology.

mechanical philosophy, especially in light of the admission of action at a distance. Attractive and repulsive forces that act a distance were deemed by Leibniz 'barbarism in physics'. But this view led him to a delicate position. On the one hand, he severely criticised his predecessors (including Descartes and Gassendi) for purging forces from nature. On the other hand, he criticised Newton and his followers for introducing sui generis forces in nature. His own supposed *via media* was that though force must be added to mass (against the Cartesians), 'that force is exercised only through an impressed impetus' (against the Newtonians). Leibniz, to be sure, was also opposed to the sui generis powers of traditional medieval-Aristotelian explanations of natural phenomena, what were deemed occult qualities, on the grounds that they 'lead us back to the kingdom of darkness'. For him the claim that all derivative forces in nature – that is, all natural forces – should be forces exerted in the collision between bodies (no matter whether they are visible or invisible), that is, forces acting by contact, was a condition of intelligibility of dynamical explanation: contact action offered an *intelligible* mechanism which underscores all natural forces. Leibniz was quick to reprimand Locke for abandoning the view that 'no body is moved except through the impulse of a body touching it'.

Leibniz's 'principles of mechanism' entailed that neither Descartes (and the Cartesians) nor Newton (and the Newtonians) offered proper mechanical explanations of natural phenomena. The culprit in Descartes's case was the lack of a notion of force (the Leibnizian *vis viva*), whereas in Newton's case it was the non-impulsive nature of gravity.

The point here is that the category of *force* was firmly introduced alongside the traditional mechanical categories (the 'most catholic principles') of *matter* and *motion*. Though Newton insisted that his concept of force was mathematical (see *Principia*, Book I, Definition VIII), he actually set it in a mechanical framework in which it was measured by the *change* in the quantity of motion it could generate. And yet mechanical interactions were enriched to include attractive and repulsive forces between particles. Mechanical explanation was taken to consist in the subsumption of phenomena under *Newton's laws*.

That's why we stressed in Section 2.1 that there is partly a terminological issue here: we might as well designate Newton's explanations in terms of forces acting at a distance 'mechanical', provided we keep in mind that we have here a new and more liberal notion of mechanism: mechanical is a system subject to Newton's laws, namely, the laws of *mechanics*.

Capitalising on Gregor Schiemann's enlightening book (2008), it can be argued that even within what we have called the mechanical conception of mechanism, there have been two distinct senses of mechanism, one wide and another narrow. The wide sense takes it that matter in motion is the ultimate cause of all natural phenomena. As such, mechanism covers everything, but its content is quite unspecific, since there is no commitment to specific laws or principles that govern the workings of the mechanism. The narrow sense of mechanism, on the other hand, has it that mechanisms are governed by the *laws of mechanics*, as enunciated paradigmatically by Newton and Lagrange. Mechanics becomes privileged because it offers universal structural principles. But then, the form of the mechanical conception of nature depends on the details of the principles of mechanics, and the content of the concept of mechanical mechanism is specified by the historical development of mechanics.

Schiemann draws an important distinction between monistic and dualistic conceptions of mechanics and, consequently, of mechanisms. On the monistic conception, there is only one fundamental mechanical category; on the dualistic conception, there are two fundamental categories. The monistic conception is further divided into two sub-categories: one takes matter to be the fundamental mechanical concept (called 'materialist' by Schiemann), while the other takes force to be the single fundamental mechanical category (called 'dynamic' by Schiemann). Huygens and Descartes had materialist conceptions of mechanical mechanism, while Leibniz had a dynamic conception. The dualist conception of mechanical mechanism admits two distinct fundamental mechanical concepts – matter *and* force. Newton was a dualist in this sense and so was Helmholtz, according to Schiemann. Helmholtz's case is particularly instructive since he proved the principle of conservation of energy. It is precisely this principle that, as we shall see in the next section, holds the key to the very possibility of a mechanical explanation of all phenomena.[2]

With the emergence of systematic theories of heat, electricity and magnetism, one of the central theoretical questions was how these were related to the theories of mechanics. In particular, did thermal, electrical and magnetic phenomena admit of *mechanical* explanations?

This question was addressed in two different ways. One, developed mostly in Britain, was by means of building of mechanical *models*. These models were meant to show (1) the realisability of the system under study

---

[2] As Schiemann (2008, 90) notes, what made the principle of conservation of energy special, at least for Helmholtz, was that energy can 'be used directly for measuring things (particularly mechanical work and heat) and their conserving properties can be examined experimentally in physical processes'.

(e.g., the electromagnetic field) by a mechanical system and (2) the possible inner structure and mechanisms by means of which the physical system under study operates. The other way was developed mostly in continental Europe and was the construction of abstract mechanical *theories* under which the phenomena under study were subsumed and explained. These theories were mechanical because they started with principles that embodied laws of mechanics and offered explanation by deductive subsumption. This tradition scorned the construction of mechanical models (especially of the wheels-and-pulleys form that many British scientists of the time were fond of). But even within this model-building tradition, especially in its mature post-Maxwellian period, mechanical models were taken to be, by and large, heuristic and illustrative devices – the focus being on the development of systematic theories (mostly based on abstract theoretical principles such as those of Lagrangian mechanics) under which the phenomena under study were subsumed and explained. Joseph Larmor (1894, 417) drew this division of labour clearly when he noticed

> [t]he division of the problem of the determination of the constitution of a partly concealed dynamical system, such as the aether, into two independent parts. The first part is the determination of some form of energy-function which will explain the recognised dynamical properties of the system, and which may be further tested by its application to the discovery of new properties. The second part is the building up in actuality or in imagination of some mechanical system which will serve as a model or illustration of a medium possessing such an energy function.

Indeed, as one of us has argued in detail elsewhere (Psillos 1999, 132–4), this distinction, in effect between an abstract mechanical (Lagrangian) theory and the various concrete configurations of matter in motion, became commonplace among the Maxwellians after James Clerk Maxwell's mature dynamical theory of electromagnetic field which rested on the general principles of Lagrangian dynamics and was independent of any particular model concerning the carrier of light waves (see Maxwell 1873, vol. 2, chapters 5–9; Klein 1972, 69–70). Maxwell made clear that the principles of Lagrangian mechanics allowed him to pursue the most general laws of behaviour of the electromagnetic field, whose constitution is unknown. He stressed:

> We know enough about electric currents to recognise, in a system of material conductors carrying currents, a dynamical system which is the seat of energy, part of which may be kinetic and part potential. The nature of the connexions of the parts of this system is unknown to us, but as we have

dynamical methods of investigation which do not require a knowledge of the mechanism of the system, we shall apply them to this case. (vol. 2, p. 213)

Hence, he formulated the kinetic and potential energies of the system in terms of electric and magnetic magnitudes and proceeded – by means of Lagrangian principles – to the derivation of the laws of motion of this system, thereby deriving the equations of the EM field (vol. 2, p. 233).[3]

2.3   Poincaré's Problem

This liberalisation of the conception of mechanical explanation, together with conceptual issues in the foundations of mechanics, brought with it the following question, which came under sharp focus towards the end of the nineteenth century: How exactly was the idea of a mechanical explanation to be rendered? The problem here was not so much related to the nature of explanation as to what principles count as *mechanical*. In 1900, Poincaré addressed the International Congress of Physics in Paris with the paper 'Relations entre la physique expérimentale et de la physique mathématique' (1900; this paper was reproduced as chapters nine and ten of his *La science et l'hypothése* of 1902). He did acknowledge that most theorists had a constant predilection for explanations borrowed from mechanics. Historically, these attempts had taken two particular forms: either they traced all phenomena back to the motion of molecules acting at a distance in accordance with central force laws or they suppressed central forces and traced all phenomena back to the contiguous actions of molecules that depart from the rectilinear path only by collisions. 'In a word', Poincaré said, 'they all [physicists] wish to bend nature into a certain form, and unless they can do this they cannot be satisfied' (ibid.). And he immediately queried: 'Is nature flexible enough for this?'

The answer is positive, but in a surprising way. Poincaré's groundbreaking contribution to this issue was the proof of a theorem that a necessary and sufficient condition for a complete mechanical explanation of a set of phenomena is that there are suitable experimental quantities that can be identified as the kinetic and the potential energy such that they

---

[3] For a brief account of Maxwell's derivation of the equations of the field, see Andrew Bork (1967). For a detailed historical account of the derivation of the equations in the symmetrical form known today, see Hunt (1991, 122–8 and 245–7).

satisfy the principle of conservation of energy.[4] Given that such energy functions can be specified, Poincaré proved that there will be *some* configuration of matter in motion (i.e., a configuration of particles with certain positions and momenta) that can underpin (or model) a set of phenomena. As he put it:

> In order to demonstrate the possibility of a mechanical explanation of electricity, we do not have to preoccupy ourselves with finding this explanation itself; it is sufficient to know the expressions of the two functions T and U which are the two parts of energy, to form with these two functions the equations of Lagrange and, afterwards, to compare these equations with the experimental laws. (1890/1901, viii)

Poincaré presented these results in a series of lectures on light and electromagnetism – delivered at the Sorbonne in 1888 and published as *Électricité et optique* in 1890 – which primarily aimed to deliver Maxwell's promise, that is, to show that electromagnetic phenomena could be subsumed under, and represented in, a suitable mechanical framework. As Poincaré put it, he aimed to show that 'Maxwell does not give a mechanical explanation of electricity and magnetism; he confines himself to showing that such an explanation is possible' (p. iv). In effect, Poincaré noted that once the first part of Larmor's foregoing division of labour is dealt with, the second part (the construction of configurations of matter in motion) takes care of itself. Maxwell's achievement, according to Poincaré, was precisely this and he 'was then certain of a mechanical explanation of electricity' (1902/1968, 224).

The irony was that Poincaré's demonstration had the following important corollary: if there is one mechanical explanation of a set of phenomena – that is, if there is a possible configuration of matter in motion that can underpin a set of phenomena – there are an *infinity* of them. And not just that. Another theorem proved by the French mathematician Gabriel Königs suggested that for any material system such that the motions of a set of masses (or material molecules) is described by a system of linear differential equations of the generalised coordinates of these masses, these differential equations (which are normally attributed to the existence of forces between the masses) would be satisfied even if one replaced all forces by a suitably chosen system of *rigid connections* between these masses. Indeed, Heinrich Hertz (1894/1955) had made use of this result to develop a system of mechanics that did away with forces altogether.

---

[4] The details of the proof (as well as further discussion of Poincaré's conception of mechanical explanation) are given in Psillos (1995).

Poincaré thought that these formal results concerning the multiplicity of mechanical configurations that could underpin a set of phenomena described by a set of differential equations were natural. They were only the mathematical counterpart of the well-known historical fact that in attempting to form potential mechanical explanations of natural phenomena, scientists had chosen several theoretical hypotheses, for example, forces acting at a distance, retarded potentials, continuous or molecular media, hypothetical fluids and so on. Poincaré was sensitive to the view that even though some of these attempts had been discredited in favour of others, more than one potential mechanical model of, say, electromagnetic phenomena was still available (see 1900, 1166–7).[5]

So, the search for a *complete* mechanical explanation of electromagnetic phenomena was heavily underdetermined by possible configurations of matter in motion. Different underlying mechanisms could all be taken to give rise to the laws of electromagnetic phenomena. By the same token, though the possibility of a mechanical explanation of electromagnetic phenomena is secured, the empirical facts alone could not dictate any choice between different mechanical configurations that satisfy the same differential equations of motion. The choice among competing underlying mechanisms (possible configurations of matter in motion) was heavily underdetermined by the empirical facts. How then can one choose between these possible mechanical configurations? How can one find the correct complete mechanical explanation of electromagnetic phenomena? For Poincaré this was a misguided question. As he said, 'The day will perhaps come when physicists will no longer concern themselves with questions which are inaccessible to positive methods and will leave them to the metaphysicians' (1902/1968, 225). His advice to his fellow scientists was to content themselves with the possibility of a mechanical explanation of all conservative phenomena and to abandon hope of finding the true mechanical configuration that underlies a particular set of phenomena. He

---

[5] The turning point in Poincaré's thinking about mechanics is in his review of Hertz's (1894) book for *Revue Générale des Sciences*. Concerning the 'classical system', which rests on Newton's laws, Poincaré agreed with Hertz that it ought to be abandoned as a foundation for mechanics (see 1897, 239). Part of the problem was that there were no adequate definitions of force and mass. But another part was that Newton's system was incomplete precisely because it passed over in silence the principle of conservation of energy (cf. 1897, 237). Like Hertz, Poincaré was more sympathetic to the 'energetic system', which was based on the principle of conservation of energy and Hamilton's principle that regulates the temporal evolution of a system (cf. 1897, 239–40). According to Poincaré (1897, 240–1) the basic advantage of the energetic system was that in a number of well-defined cases, the principle of conservation of energy and the subsequent Lagrangian equations of motion could give a full description of the laws of motion of a system.

stressed: 'We ought therefore to set limits to our ambition. Let us not seek to formulate a mechanical explanation; let us be content to show that we can always find one if we wish. In this we have succeeded' (1900, 1173).

According to Poincaré, the search for mechanical explanation (i.e., for a configuration of matter in motion) of a set of phenomena is of little value not just because this search is massively underdetermined by the phenomena under study but mainly because this search sets the wrong target. What matters, for Poincaré, is not the search of mechanism per se, but rather the search for *unity* of the phenomena under laws of conservation. Understanding is promoted by the unification of the phenomena and not by finding mechanical mechanisms that bring them about. As he said, 'The end we seek ... is not the mechanism. The true and only aim is unity' (p. 1173).

One may question the status of the law of conservation of energy as a mechanical principle. But that's beside the point. For the point is precisely that there is no fixed characterisation of what counts as mechanical. It may well be that Poincaré's notion of mechanical explanation is too wide from the point of view of physical theory, since it hardly excludes any phenomena from being subject to mechanical explanation. Still, and this is quite important, it does block certain versions of vitalism that stipulate new kinds of forces. As is well known, in the twentieth century, the search for mechanisms and mechanical explanations was taken to be a weapon against vitalism. One key problem with vitalist explanations (at least of the sort that C. D. Broad has dubbed *Substantial Vitalism*) is that they are in conflict with the principle of conservation of energy, and in *this* sense, they cannot be cast, even in principle, as mechanical explanations.

The significance of Poincaré's problem for the mechanical conception of mechanism can hardly be overestimated. But we should be careful to note exactly what this problem is. It is not that mechanical mechanisms are unavailable or non-existent. It is not that nature is *not* mechanical. Hence, it is not that mechanical explanation – that is, explanation in terms of mechanical mechanisms – is impossible. On the contrary, Poincaré has secured its very possibility, thereby securing, as it were, the victory of traditional mechanical philosophy over Aristotelianism. Rather, the problem for the mechanical conception of mechanism that Poincaré has identified is that mechanical mechanisms are *too* easy to get, provided nature is conservative. Under certain plausible assumptions that involve the principle of conservation of energy, the call for mechanical explanation is so readily satisfiable that it ceases to be genuinely informative.

2.4 Quasi-Mechanical Mechanisms

Ewing (1969, 216) drew a distinction between two conceptions of mechanical necessity in Kant's *Third Critique*. The first is related to what we have called the mechanical conception of mechanism: a determination of the properties of a whole by reference to matter in motion, and in particular by the mechanical properties of its parts and the mechanical laws they obey. The second, which Ewing calls 'quasi-mechanical', is still a determination of the properties of the whole by reference to the properties of its parts, but with no particular reference to mechanical properties and laws. This quasi-mechanical conception of mechanism is broader than the mechanical conception since there is no demand that the laws that govern the behaviour of the parts, or the properties of these parts, are mechanical – at least in the strict sense associated with the mechanical conception.

Peter McLaughlin (1990) has developed a similar account of Kant's conception of mechanical explanation, according to which the mechanism of nature is a form of causation, whose differentia is that it takes it that the whole is determined by its parts. Thus understood, a mechanical explanation is a kind of decompositional (or modular) explanation: an explanation of a whole in terms of its parts, their properties and their interactions. McLaughlin bases his account on the following point made by Kant in his *Critique of Judgement* (1790/2008, 408): 'Now where we consider a material whole, and regard it as in point of form a product resulting from the parts and their powers and capacities of self-integration (including as parts any foreign material introduced by the co-operative action of the original parts), what we represent to ourselves in this way is a mechanical generation of the whole.' Accordingly, what renders a structure a mechanism is the fact that it possesses a reductive unity: its behaviour is determined by the properties its parts have 'on their own, that is independently of the whole' (McLaughlin 1990, 153).

This is not the place to discuss in any detail whether this was indeed Kant's own conception.[6] The key point is that *if* this conception is viable at all (and, as the current mechanistic turn demonstrates, it *is*), then the concept of mechanism is not tied to mechanics, nor to the operation of specifically mechanical laws, nor to the ultimate determination of the

---

[6] There are competing views on this. Hannah Ginsborg takes it that Kant's conception of mechanism is closely tied to his account of forces and mechanical laws. For her, according to Kant, 'we explain something mechanically when we explain its production as a result of the unaided powers of matter as such' (2004, 42). For an attempted synthesis of Ginsborg's and McLaughlin's views, see Breitenbach (2006).

behaviour of mechanism by reference to mechanical properties and inter-actions. Rather, the mechanism is *any* complex entity that exhibits reductive stability and unity in the sense that its behaviour is determined by the behaviour of its parts.

Kant, to be sure, contrasted mechanical explanation to teleological explanation. In its famous antinomy of the teleological power of judgement, he contrasted organisms to mechanisms. *Qua* material things, organisms (like all material things) should be generated and governed by merely mechanical laws. And yet, some material things (*qua* organisms, and hence natural purposes, as Kant put it) 'cannot be judged as possible according to merely mechanical laws (judging them requires an entirely different law of causality, namely that of final causes)' (1790/2008, 387). The defining characteristics of an organism – that is of a non-mechanism – are two: (1) the whole precedes its parts and, ultimately, determines them; and (2) the parts are in reciprocal relations of cause and effect. Famously, Kant claimed that the very idea of non-mechanism (organism) is regulative and not constitutive – we have the right to proceed *as if* there were organisms (non-mechanisms) but this is not something that can be known or proved, though Kant did think that this regulative principle is a *safe* presupposition, not liable to refutation by the progress of science.

This contrast of mechanism and non-mechanism suggests that the key feature of mechanism – what really sets it apart from organism – is the priority of the parts over the whole in the constitution of the mechanism and the determination of its behaviour.[7] It is also worth noting that it is precisely this contrast that C. D. Broad (1925) had in mind in his own critique of mechanism.

Broad mounted an attack on what he called 'the ideal of Pure Mechanism'. This is an extreme and purified version of what we have called the mechanical conception of mechanism. Broad's Pure Mechanism is a world view, which he (1925, 45) characterises thus:

---

[7] Ginsborg (2004) takes it that *qua* natural purposes, organisms are non-machine-like (and hence mechanically inexplicable) in the sense that 'they are not assemblages of independent parts, but that they are instead composed of parts which depend for their existence on one another, so that the organism as a whole both produces and is produced by its own parts, and is thus in Kant's words 'cause and effect of itself' (p. 46). This way to read Kant's account of organism distinguishes it from mechanism in two senses. (1) Organism cannot be explained in terms of the powers of matter as such, and (2) organism is such that its parts depend on the whole and cannot 'exist independently of the whole to which they belong' (p. 47). Hence, what renders mechanism distinctive is precisely the fact that its unity and behaviour are determined by its parts, as they are independent of their presence in the whole. For a useful attempt to synthesise Kant's antinomy in the light of modern evolutionary biology, see Walsh (2006).

The essence of Pure Mechanism is

a.   a single kind of stuff, all of whose parts are exactly alike except for differences of position and motion;

b.   a single fundamental kind of change, viz, change of position . . .;

c.   a single elementary causal law, according to which particles influence each other by pairs; and

d.   a single and simple principle of composition, according to which the behaviour of any aggregate of particles, or the influence of any one aggregate on any other, follows in a uniform way from the mutual influences of the constituent particles taken by pairs.

The gist of Pure Mechanism is that it is an ontically reductive thesis and in particular a reductive thesis with a very austere reductive basis of a single kind of fundamental particle, a single kind of change and a single causal law governing the interaction of the fundamental particles. Broad contrasted this view with two others. The first is what he called *Emergent Vitalism*. This is the view that living organisms and their behaviour cannot be fully and exhaustively determined by the properties and behaviour of their component parts, as these would be captured by the ideal of Pure Mechanism. Emergent Vitalism is also opposed to a view we have already noted in Section 2.3, namely, Substantial Vitalism: that living organisms are set apart from mechanism by an extra element (a kind of life-conferring force) that they share while pure mechanisms do not. In denying Substantial Vitalism, Emergent Vitalism puts emphasis on the structural arrangement of the whole vis-à-vis its parts and on the interaction among the parts when they are put together in a whole. A certain whole W may consist of constituents A, B, C placed in a certain relation R(A, B, C). There is emergence – emergent properties – when A, B, C cannot determine, even in principle, the properties of R(A, B, C).

Broad (1925, 61) put this point in terms of the lack of an in-principle deducibility of the properties of R(A, B, C) 'from the most complete knowledge of the properties of A, B, and C in isolation or in other wholes which are not of the form R(A, B, C)'. This way to put the matter might be unfortunate, since what really matters is the metaphysical determination (or its lack thereof) of the whole by its parts and not deducibility per se – which is dependent on the epistemic situation we might happen to be in. But what matters for our purposes is Broad's thought that the denial of Pure Mechanism need not lead to the admission of spooky forces and mysterious powers, associated with Substantial Vitalism.

Still, our main concern here is not the opposition of Pure Mechanism to Emergent Vitalism, but rather its opposition to what Broad rightly took to be a milder form of mechanism. This form, which Broad associated with what he called *Biological Mechanism*, is committed to the view that the behaviour of a whole (and of a living body in particular) is determined by its constituents, their properties and the laws they obey, but relies on a broader conception of what counts as a constituent and what laws are admissible. As Broad put it:

> Probably all that he [a biologist who calls himself a 'Mechanist'] wishes to assert is that a living body is composed only of constituents which do or might occur in non-living bodies, and that its characteristic behaviour is wholly deducible from its structure and components and from the chemical, physical and dynamical laws which these materials would obey if they were isolated or were in non-living combinations. Whether the apparently different kinds of chemical substance are really just so many different configurations of a single kind of particles, and whether the chemical and physical laws are just the compounded results of the action of a number of similar particles obeying a single elementary law and a single principle of composition, he is not compelled as a biologist to decide. (p. 46)

This is, clearly, what we have called a quasi-mechanical conception of mechanism, and as Broad rightly notes, this kind of conception is enough to set the mechanist biologist apart from the emergent vitalist. The controversy need not be put, nor is it useful to be put, in terms of the ideal of Pure Mechanism.[8]

Enough has been said, we hope, to persuade the reader that there is a distinct quasi-mechanical idea of mechanism, which – to recapitulate – proclaims a form of determination of a whole by its parts, their properties and interactions, as these would occur independently of their presence in the whole. With this in mind, let us now see what the key problem of this quasi-mechanical conception of mechanism is.

2.5 Hegel's Problem

Seeing the heading above, the reader might wonder: Has something gone awfully wrong? What can Hegel possibly have to do with mechanisms and

---

[8] In his very useful (2005) paper, Garland Allen notes that 'operative, or explanatory mechanism refers to a step-by-step description or explanation of how components in a system interact to yield a particular outcome' (p. 261). He contrasts this with what he calls 'philosophical mechanism', which he takes to assert that living things are material entities. He then offers an instructive historical account of approaches to biological mechanism in the early twentieth century (and their opposition to vitalism), emphasising that 'the form that Mechanistic thinking took in the early twentieth century ... differed from earlier (eighteenth- and nineteenth-century) mechanistic traditions. It was physico-chemical not merely mechanical' (p. 280).

contemporary mechanistic explanation in the sciences? Well, it turns out not only that there is a quite different conception of mechanism that goes back (at least) to Hegel, but that he identified a problem that all accounts of mechanism (especially those mechanisms which we shall call mechanisms-for) should meet.

Long before Poincaré's critique of mechanical mechanism, Hegel had, in his *Science of Logic*, attacked the idea that all explanation must be mechanical. According to James Kreines (2004), Hegel argued that making mechanism an absolute category – applicable to everything – obscures the distinction between explanation and description and hence undermines itself.

Hegel's writings on mechanism are rather cryptic (and, perhaps, obscure). Essentially, he took the characteristic of mechanism to be that it possesses only an external unity. Its constituents (the objects that constitute it) retain their independence and self-determination, although they are parts of the mechanism. As he put it in his *The Encyclopaedia Logic* (1832/1991, 278), 'the relation of mechanical objects to one another is, to start with, only an external one, a relation in which the objects that are related to one another retain the semblance of independence'. And in his *Science of Logic* (2002, 711) he stressed: 'This is what constitutes the character of *mechanism*, namely, that whatever relation obtains between the things combined, this relation is one *extraneous* to them that does not concern their nature at all, and even if it is accompanied by a semblance of unity it remains nothing more *than composition, mixture, aggregation* and the like.' The determinant of the unity of a mechanism or, as Hegel put it, 'the *form* that constitutes [its] difference and combines [it] into a unity' is 'an external, indifferent one; whether it be a *mixture*, or again an *order*, a certain *arrangement* of parts and sides, all these are combinations that are indifferent to what is so related' (p. 713). And elsewhere, he stressed that, being external, the unity of the mechanism 'is essentially one in which no *self-determination* is manifested' (p. 734).

On Kreines's reading of Hegel's critique of mechanism, Hegel raised a perfectly sensible and quite forceful objection to the view that *all* explanation is mechanical explanation, that the only mode of explanation is mechanical and that to explain X is to offer a mechanical explanation of it.

Hegel's argument against the idea of mechanism – *qua* an all-encompassing explanatory concept – goes like the following. Mechanistic explanation proceeds in terms of breaking an object down to its parts and of showing its dependence on them and their properties and relations. Explanation, then, amounts to a certain decomposition of the explanandum, namely, of a composite object whose behaviour is the result of the properties of, and interactions among, its parts. But there are indefinitely many ways to decompose something to parts and to relate it and its behaviour to them. For the call for explanation to have any bite at all, there must be some principled distinction between those decompositions that are merely descriptions of the explanandum and those decompositions that are genuinely explanatory. In particular, some decomposition – that which offers the mechanical *explanation* – must be privileged over the others, which might well reflect only pragmatic criteria or subjective interests. But how is this distinction to be drawn within the view that all explanation is mechanical? If all explanation is indeed mechanical, and if mechanical explanation amounts to decomposition, no line can be drawn between explanation and description – no particular way to decompose the explanandum is privileged over the others by being mechanical – mechanical as opposed to what? All decompositions will be equally mechanical and equally arbitrary. Hence, there will be no difference between explanation and description.

Hegel was pushing this line of argument in order to promote his own organic view of nature and, in particular, to reinstate a teleological kind of explanation – one that explains the unity of a composite object in terms of its internal purposeful activity.[9] But the point he makes is very general. In essence, Hegel's problem is that something external to the mechanism (considered as an aggregate of parts) is necessary to understand how mechanistic explanation is possible. His general point is that the unity of a mechanism is not just a matter of arranging a set of elements into a whole; nor is it just a matter of listing their properties and mutual relations. Nor is it determined by the parts of the mechanism, as they are independent of their occurrence within the mechanism. There are indefinitely many ways to arrange parts into wholes or to decompose wholes into parts. Most of them will be arbitrary since they will *not* be explanatorily relevant. The unity of the mechanism comes from something external to it, namely, from its function – from what it is meant to be a mechanism *for*. The function that a mechanism performs is something external to the description of the mechanism. It is the function that fixes a criterion of explanatory relevance. Some descriptions of the mechanism are explanatorily relevant while others are not because the former and not the latter explain how the mechanism performs a certain function.

---

[9] For an informative and intelligible account of Hegel's organic world view, see Beiser (2005, chapter 4).

Let us illustrate this point with a couple of examples. Consider a toilet flush – a very simple mechanism indeed. What confers unity to it *qua* mechanism is the function it performs. As a complex entity, it can be decomposed into elements in indefinitely many different ways. Actually, in all probability, there is, in principle, a description of it in terms of the interactions of the atoms of water and their collisions with the walls of the tank and so on. What fixes the explanatorily relevant description is surely the function it performs. Or consider a telephone conversation through which some information is passed over from one end to the other – a very simple social mechanism. What confers unity to it *qua* mechanism is its function, namely, to transfer information between two ends. In all probability, there is a description of this mechanism in terms of the interactions of sound waves, collisions of particles, triggering of nerve-endings and so on. But this description is explanatorily irrelevant when it comes to explaining how this simple social mechanism performs its function. Notice that a point brought out by these examples, and certainly a point that Hegel had in mind, is that the truth of a description (supposing that it is to be had) does not necessarily render this description explanatorily relevant.

We could sum up Hegel's problem like this: first the function, then the mechanism.[10] The functional unity of the mechanism determines, ultimately, which of the many properties that the constituents of the mechanism have are relevant to the explanation of the performance and function of the whole. Hegel (1832/1991, 275) did think that mechanism is a form of objectivity, claimed that it is applicable to areas other than 'the special physical department from which it derives its name' but denied that it is an 'absolute category' that is constitutive of 'rational cognition in general'.

2.6   Bringing Together the Two Problems

Qua thinkers, Hegel and Poincaré were as different as chalk and cheese. Yet they both point – with different philosophical arguments – towards a decline of the mechanistic world view. It's not that there are no mechanisms. Actually, mechanisms, in the broad sense of stable arrangements of matter in motion, are ubiquitous. But it does not follow from this that nature has a definite mechanical structure (or, if that's too strong, that we cannot know which definite mechanical structure is the one actually characterising nature). This is, in essence, Poincaré's problem. How the

---

[10] This is indeed something that many new mechanists have come to accept.

mechanisms are individuated is a matter external to them: what counts as a mechanism, where it starts and where it stops, what kind of parts are salient and what kind of properties are relevant depend on the function they are meant to perform. The unity of the mechanisms is not intrinsic but extrinsic to them. This is, in essence, Hegel's problem. But even after a function has been determined, there are indefinitely many ways to configure mechanical mechanisms that perform it, that is, to offer a mechanical model (a configuration of matter in motion) that performs it. This is a corollary of Poincaré's problem.[11] Nature, even if it is mechanical, does not fix the boundaries of mechanisms. When it comes to the search for mechanisms, *anything* can count as a quasi-mechanism provided it performs a function that it is meant to explain. This is a corollary of Hegel's problem.

Does the search for mechanisms improve understanding? The answer is unequivocally positive. The description of a mechanism is a theoretical description and, as such, tells a story as to how the phenomenon under study is brought about – if the story is true, our understanding of nature is enhanced. Insofar as mechanisms are taken to be functionally individuated stable explanatory structures (whose exact content and scope may well vary with our best conception of the world) which enhance our understanding of how some effects are brought about or are the realisers of certain functions, they can play a useful role in the toolkit of explanation.

What does all this imply about the ontic status of mechanisms? Given that the world is governed by conservation laws, it allows descriptions of configurations of matter in motion subject to the laws of mechanics; that is, it admits of (a multitude of) mechanical mechanisms. But that's all spoils to the victor, the latter being the fact of the law of conservation of energy. Even then, however, mechanisms are functionally individuated; the spoils do not come to much – there is more than one way to peel an orange!

---

[11] Hegel was confident that 'not even the phenomena and processes of the physical domain in the narrower sense of the word (such as the phenomena of light, heat, magnetism, and electricity, for instance) can be explained in a merely mechanical way (i.e., through pressure, collision, displacement of parts and the like' (1832/1991, 195). Poincaré proved him wrong on this.

PART II

Causation and Mechanism

CHAPTER 3

Mechanisms in Scientific Practice
The Case of Apoptosis

3.1 Preliminaries

Talk of mechanisms is widespread within the life sciences. Pathologists talk about 'mechanisms of disease' (Lakhani et al. 2009) and pharmacologists about the 'mechanisms of action' of drugs (Rang et al. 2016). Molecular biologists talk about the 'fundamental mechanisms of life' (Alberts et al. 2014, 22), such as DNA replication and protein synthesis; developmental biologists talk about 'genetic mechanisms of animal development' (Alberts et al. 2014, 39), 'mechanisms for specifying the germ layers' (Wolpert et al. 2002, 89) and 'mechanisms of axis determination' (Wolpert et al. 2002, 143); and there exist 'morphogenetic mechanisms' (Wolpert et al. 2002, 254) and the 'mechanism of programmed cell death' (Slack 2005, 214).

Searching for mechanisms and mechanical explanations has been viewed as a main aim of life sciences, and science in general (see Craver & Tabery 2015). There exists a widespread consensus among philosophers of science that an adequate philosophical account of the practice of current sciences must be structured around this basic notion (see Glennan & Illari 2018a). However, philosophers have not yet reached a consensus about what a mechanism is. This lack of a generally accepted account may seem surprising to the outsider, given the prominence of the concept in scientific practice.

3.2 The Case of Apoptosis

3.2.1 Why Apoptosis?

If we want to understand what the things that scientists identify as mechanisms are, we had better examine how mechanisms are identified in practice. So, we take it that the best argument for CM is that precisely this conception of mechanism is the one that we find in biological (as well as in other scientific) contexts where the language of mechanisms is used. In order to substantiate this argument from scientific practice, we will examine in depth a central example of a biological mechanism, the mechanism of cell death known as apoptosis. We will use apoptosis as a benchmark to develop a bottom-up argument (i.e., one based on a case study of a prominent case of a biological mechanism) in favour of the adequacy of CM. Our discussion in this chapter will establish that CM is a typical use of 'mechanism' in life sciences. In the next chapter and the chapters to follow, we will generalise CM and apply it to other cases.

There are three main reasons why apoptosis is a particularly good example for our purposes. The term 'apoptosis' was introduced in the biological literature by John F. R. Kerr, Andrew H. Wyllie and Alastair R. Currie in a seminal 1972 paper as the name of a newly discovered mechanism, and in particular of 'a hitherto little recognised mechanism of controlled cell deletion' (Kerr et al. 1972, 239). It is thus a particularly instructive case that can be used to draw lessons about how a new mechanism is identified in biology, and hence to provide insights about what exactly a mechanism in life sciences is taken to be. Subsequent research has identified apoptosis as a ubiquitous mechanism for the regulation of cell populations with important clinical relevance and has offered various levels of description of its workings. As we will see, this provides further insights into the nature of mechanisms in biology. Last, the study of apoptosis (and of mechanisms of programmed and physiological cell death in general) transcends particular biological fields and has involved cytologists, developmental biologists, pathologists and molecular biologists, among others. Because of its broad role, this case offers a nice test case of the concept of mechanism as it is used in the life sciences. Apoptosis is described in all these biological disciplines as a 'mechanism' of cell death. The common concept of mechanism at play here, we will argue, is that a mechanism is just a causal pathway. The case of apoptosis will then provide strong reasons for the view that CM is the common denominator of (at least) many uses of the concept of mechanism in biology.

3.2.2 Early History of Cell Death

The history of cell death can be traced back to the mid-nineteenth century, when biologists were already aware that there exist processes that lead to the death of cells (see Clarke & Clarke 1996). Cytologists of the time, such as Walther Flemming in 1885, had even observed cells undergoing what we now consider as *apoptosis*. However, in the following decades and during the 1960s and even later, there was not much interest from biologists in cell death. To explain why this was so, Richard A. Lockshin notes that biologists at that time 'tended to think that death was accidental and that mitosis was the active homeostatic process' – cell death was not yet viewed as a 'biological process' (Lockshin 2008, 1092).

In the 1960s, new technological developments (e.g., electron microscopy, improved histological techniques) started a new era in the study of cell death. In 1964, Lockshin and Carroll Williams published a paper with the title 'Programmed Cell Death' (Lockshin & Williams 1964). According to Lockshin and Zahra Zakeri, this new expression ('programmed cell death') was meant to capture the fact that 'cells followed a sequence of controlled (and thereby implicitly genetic) steps towards their own destruction' (Lockshin & Zakeri 2001, 546).

That such a controlled process existed was well known to developmental biologists. John W. Saunders had already noticed that there exist 'reproducible patterns of cell death in chick embryos' (Lockshin & Zakeri 2001, 546) and the same was the case concerning metamorphic cell deaths in insects. As he put it, 'abundant death, often cataclysmic in its onslaught, is a part of early development in many animals; it is the usual method of eliminating organs and tissues that are useful only during embryonic or larval life or that are but phylogenetic vestiges' (Saunders 1966, 154). Characteristically, Saunders wrote about a 'death clock' intrinsic in cells: cells that normally die during chick development would also die 'on schedule' in culture. If, however, they were transplanted to a different area of other chicks, they survived. It was evident from this that cell death is a controlled and regulated process.

To be sure, the idea of 'programmed cell death' was metaphorical. As Lockshin explains in a recent review on the history of the subject, it was

> a felicitous turn of phrase designed to exploit the trendiness of the then-nascent computer era. The intent was to focus attention on what was relatively obvious: that cell deaths in developing and metamorphosing animals occurred at predictable developmental stages and in specific locations. They must be 'programmed' into the genetics of the organisms, in the same sense that the differentiation and growth of an organ, tissue, structure, or pigment would be considered to be fundamentally determined by the interplay of specific genes. (Lockshin 2016, 10–11)

3.2.3   Identifying Apoptosis

John F. R. Kerr, who had been working on the processes of cell death since the 1960s, notes that at the time most researchers were 'equating cell death with cell degeneration' (Kerr 2002, 472). So, an early hypothesis was that cell death was the result of damaged lysosomes, which were viewed as 'suicide bags'. Kerr, however, had discovered a certain type of cell death that was 'non-degenerative in nature' (2002, 472) – he first named it 'shrinkage necrosis' (Kerr 1971). This was no accident; the initial thought was that this process was a type of *necrosis*. But soon Kerr noted that it was a different kind of mechanism – what he and his collaborators called 'apoptosis'.

Shrinkage necrosis was identified by studying ischaemic liver injury. It was a type of cell death that differed from classical necrosis both morphologically, in that it involved scattered cells that were converted in small round bodies, rather than groups of cells as in classical necrosis, and also chemically, in that during shrinkage necrosis lysosomes were preserved, again in contrast to classical necrosis where they ruptured. In his 1971 paper, Kerr concluded: 'Shrinkage necrosis is a distinct and important type of cell death, which has received relatively little attention in the past. It probably results from noxious stimuli that are insufficiently severe to produce coagulative necrosis' (p. 19).

The concept of apoptosis was introduced in a seminal paper in 1972 by Kerr, Wyllie and Currie as 'a hitherto little recognized mechanism of controlled cell deletion' (1972, 239). What did Kerr et al. do to identify the mechanism of apoptosis? They described it in the language of theory as a 'vital biological phenomenon', which is 'complementary to mitosis in the regulation of animal cell populations' (p. 241). This was mainly a description of the specific causal pathway of the deletion of 'scattered single cells' (p. 241). (See Figure 3.1.)

There are two main stages in the apoptotic process as shown by electron microscopy: first, so-called apoptotic bodies are formed; second, apoptotic bodies are phagocytosed and degraded by other cells. Apoptotic bodies are small, spherical, membrane-bound structures that contain condensed, but otherwise intact and functional, cell organelles and fragments of nuclei. During the formation of apoptotic bodies the nucleus and the cytoplasm condense, and nucleus fragments and protuberances are formed on the surface of the cell. The cell then breaks apart and from the protuberances the apoptotic bodies are formed. Within the cells that phagocytose them, apoptotic bodies show changes that are 'very similar to ischaemic coagulative necrosis'. But, in contrast to coagulative necrosis, the absence of inflammation in apoptosis results in a process of cell death with minimal disruption of the tissue. So, apoptosis is distinct from necrosis and, as Kerr et al. put it, 'is well suited to a role in tissue homoeostasis' (p. 250).

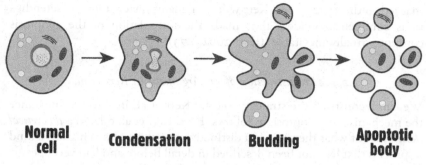

Normal cell **Condensation** **Budding** **Apoptotic body**

Figure 3.1 Apoptosis as described by Kerr et al. (1972).

Crucially, the morphological changes that occur during apoptosis were 'essentially the same' (p. 244) in various types of circumstances studied by the authors, both physiological and pathological. Establishing this point was important, as a mistake that had previously been made in various cases was to confuse apoptotic bodies that have been phagocytosed with, for example, autophagic vacuoles (which often appear similar under the electron microscope). For example, apoptosis was found by Kerr and his collaborators to occur spontaneously in both treated and untreated malignant tumours and was involved in cases of pathological atrophy but also of normal involution of tissues, in normal development (e.g., during the development of digits) and in general in cellular turnover in normal adults. It was then taken to be a ubiquitous mechanism: 'a distinctive morphological process ... which plays a complementary but opposite role to mitosis in the regulation of animal cell populations' (pp. 255–6).

By describing the morphological pattern (causal pathway) of the apoptotic process and the various circumstances where it occurs, the authors reach the crucial conclusion that apoptosis has a regulatory role within the organism: it 'subserves a general homeostatic function' (Kerr 2002, 471). This regulatory role is the most important difference with coagulative necrosis, the classical type of cell death that had already been described morphologically in detail and which Kerr et al. contrasted to apoptosis. Coagulative necrosis does not regulate cell populations, as it is brought about when homeostatic mechanisms are irreversibly disturbed and is always caused by 'noxious stimuli' (Kerr et al. 1972, 239), whereas apoptosis is caused by both pathological and physiological stimuli. The triggering of apoptosis by *physiological* stimuli is the reason why apoptosis is considered of great importance by the authors. Summing up the events

that led to the 1972 paper, Kerr writes in a later review that '[a] serendipitous confluence of ideas thus made the formulation of the apoptosis
concept virtually inevitable' (Kerr 2002, 473).

3.2.4   Reconstruction of the Argument in Kerr et al.

We can reconstruct the strategy used by Kerr et al. in order to introduce
the mechanism of apoptosis as follows. First, Kerr et al. offered a *theoretical
description* of what they saw as a distinctive causal process. This was a kind
of process that had not been described in detail before and had very specific
morphological features. A main aim of their paper was to describe those
features. Electron microscopy revealed that this process involved the
following morphological changes: condensation of cytoplasm and nucleus,
fragmentation of the nucleus, formation of protuberances on the surface of
the cell, subsequent breaking of the cell and formation of spherical
structures that are membrane-bound and contain cell organelles and nuclei
fragments that are condensed but otherwise functional (named 'apoptotic
bodies' by Kerr et al.) and finally the phagocytation and degradation of
apoptotic bodies by other cells.

Second, the authors specified the *ubiquitous character* of the new process: the morphological changes associated with it are 'essentially the same'
in many circumstances, both physiological and pathological (e.g., in
malignant tumours, in cases of pathological atrophy, in normal development and in cellular turnover in normal adults).

Third, they noted a distinctive feature concerning the new process,
namely, its *non-disruptive* nature. In particular, whereas it was a process
resulting in the death of the cell, it did not produce inflammation. This
enabled them to discriminate it from necrosis, the process that results in
the death of the cell following a toxic stimulus. However, apoptosis was
triggered also by physiological stimuli. It thus seemed to have a specific
role within the organism. This immediately gives rise to the question:
What is a main function of apoptosis?

In order to identify such a function, Kerr et al. noted that a particular
kind of process has to exist, namely, some form of 'physiological cell death'
that is at work balancing divisions in cell populations (1972, 239).
However, necrosis, due to its disruptive nature, cannot play that role.
Apoptosis, however, precisely because it (1) is non-disruptive, (2) can be
triggered by physiological stimuli and (3) is ubiquitous, is a particularly
well-suited candidate to play that role. So, apoptosis 'is well suited to a role
in tissue homoeostasis, since it can result in extensive deletion of cells with

little tissue disruption' (p. 250). The conclusion, then, is that apoptosis plays a crucial regulatory role in tissue homeostasis.

To sum up, Kerr et al. argued as follows. Since apoptosis is

i)   a distinctive morphological process
ii)   ubiquitous
iii) non-disruptive, in contrast to necrosis
iv)   triggered by physiological stimuli; and given that
v)   a form of 'physiological cell death' that is at work balancing divisions in cell populations must exist; and since
vi)   necrosis cannot play that role, while apoptosis is well suited to play that role; therefore,
vii) apoptosis is the mechanism with exactly this regulatory role across animals.

3.2.5   Extraction of Key Features

We can now extract from this case some salient features that are sufficient for the introduction and the characterisation of a new mechanism in biology. First of all, mechanisms are processes or, as we prefer to call them, borrowing a terminology widespread in biology, *causal pathways*. So, the first task that Kerr et al. had to accomplish in order to introduce the mechanism of apoptosis was to offer a theoretical description of a certain process, which is extended over time and is characterised by specific features. The process can be seen as *causal* as it is characterised by a regular sequence of causal steps and difference-making relations among its constituents: the formation of apoptotic bodies is dependent on the fragmentation of the nucleus, which depends on the condensation of cytoplasm and nucleus. The recurrence of this succession of stages under a variety of conditions offered evidence that this is indeed a causal sequence, with a possible genetic basis. As Kerr et al. put it, '[t]he ultrastructural features of apoptosis and its initiation and inhibition by a variety of environmental stimuli suggest to us that it is an active, controlled process' (p. 256).

Second, the description of the mechanism of controlled cell deletion makes clear that the full knowledge of the causal pathway is not necessary for the identification of the mechanism, as the identification of this new type of pathway did not require a full understanding (and hence a full description) of its workings. So, Kerr et al. noted that the mechanism of condensation, a main stage in the apoptotic process, was unknown. As they put it, 'the condensation is presumably a consequence of the extrusion of water, but its mechanism is still unknown' (p. 244). Additionally, in 1972 there was a lot that was not yet known 'of the factors that initiate apoptosis or of the nature of the cellular mechanisms activated before the appearance of the characteristic morphological changes' (p. 255). And of course, nothing was yet known of the biochemical processes underlying apoptosis or its genetic basis. Nevertheless, enough of its causal pathway was known to conclude that it is an 'active, controlled process' (p. 256). This shows that it is one thing to identify a mechanism for a certain phenomenon and it is quite another thing to acquire a full description of its workings. Hence, even with limited knowledge about the various causal details, it is possible to identify and initially describe a potentially important new causal pathway.

Descriptions of a mechanism can be made richer by offering more detailed characterisations. This can involve only the 'horizontal' dimension, as we may call it, of a process, for example, offering further cytological details of apoptosis; more interestingly, it can involve the 'vertical' dimension. Describing the 'mechanisms of condensation', for example, can be done at a cytological level by offering further cytological details, perhaps more fine-grained ones, or by offering details in the vertical dimension, for example, by giving a biochemical description of what is going on. Descriptions of biological mechanisms then are always couched in theoretical language and can be enriched in various ways. In particular, there can be alternative theoretical descriptions of the same mechanism; for example, we now have a cytological and a biochemical description of the apoptotic pathway (see below). Hence, at a minimum, a mechanism is *a certain theory-described causal pathway.*

Third, the regulatory role of apoptosis was crucial for its identification as a distinctive kind of mechanism. As we saw, Kerr himself in earlier work had already observed the process that was to be called apoptosis, but he did not regard it as a new kind of mechanism. Instead, he viewed it as a type of necrosis, that is, as a type of pathological cell death, that was 'non-degenerative in nature' compared with classical necrosis (Kerr 2002, 472). But shrinkage necrosis was not viewed as a new kind of mechanism.

It was the fact that apoptosis seemed to have a basic regulatory function within the organism that led, in the 1972 paper, to its identification as a new kind of mechanism. There, the apoptotic process is not taken to be a type of necrosis anymore; it is contrasted with necrosis and constitutes a new and very important kind of regulatory mechanism of cell death. Whereas the various microscopic observations described in the paper are used to establish a new causal pathway, the regulatory argument offered by

Kerr establishes that the new causal pathway constitutes a new kind of mechanism, thereby introducing a new taxonomy of types of cell death.

So, the function of a particular theoretically described process, that is, its role within the organism, is important when introducing a new biological mechanism.[1] The function serves to further distinguish the process from other similar processes, and it is crucial for establishing not only a distinct but also a new kind of mechanism. So, in our example, both necrosis and apoptosis are mechanisms of cell death. But they are distinct mechanisms since, first, they are characterised by different cytological features, and, second, they play different roles within the organism.[2]

The lesson from our discussion so far: what the identification of a new mechanism in the life sciences requires, as evidenced by the case study of apoptosis, is *a theoretical description of a causal pathway with a certain function.*

3.2.6 Apoptosis after 1972

By the mid 1970s it had already been recognised that 'cell death was as much a part of cell biology as mitosis, extension of an axon, the enzymatic sequence of glycolysis, or secretion' (Lockshin & Zakeri 2001, 547). This was a crucial conceptual breakthrough compared with the older way of thinking about cell death; before the 1970s cell death had not been seen as a phenomenon on a par with mitosis or glycolysis, which constitute fundamental biological mechanisms. However, the important point here is that even after the introduction and the wide recognition of apoptosis in the 1970s, research on apoptosis did not gain the importance it has today.

In the 1990s, however, the field was transformed into a fundamental research area within the life sciences. A major reason for this transformation was the realisation of its central role in many functions within the organism and the fact that the mechanism was understood to such an extent that it was seen as a phenomenon that was medically central. This reflects, then, a second major breakthrough in the history of apoptosis. By

---

[1] When we talk about biological functions of processes of causal pathways here and elsewhere, we mean the role a process plays in an organism that is the result of selection (cf. Millikan 1984; Neander 1991); apoptosis has been selected for regulating cell populations and has thus a homeostatic function, while necrosis does not. Artifacts, as products of design, can also have functions.

[2] New mechanists take mechanisms to be always mechanisms *for* a phenomenon, but typically this is not taken to entail that the phenomenon is some function. See Garson (2018) for an overview of the discussion. We agree that not all mechanisms have functions.

the mid 1990s, cell death 'was recognized as an interesting and biological event'; it was seen not just as 'an incidental part of life' but as 'a highly controlled and medically important element of existence' (Lockshin & Zakeri 2001, 545).

More specifically, there are three main reasons that can explain why the field of apoptosis was transformed from a modest topic to a central field of biological research. First, it was discovered that apoptosis was much more common than was initially thought by the development of techniques that made it easier to identify instances of apoptosis. Second, conserved genes that control cell death were identified, starting with the genes that determine the developmental pathway for programmed cell death in *Caenorhabditis elegans* (see Ellis & Horvitz 1986). Unravelling the molecular genetic mechanisms regulating cell death by using *C. elegans* as a model that began in late 1970s was a breakthrough in the study of regulated cell death. It was thus established that cell death was genetically based, and various important genes that are involved in the regulation of apoptosis were identified (e.g., bcl-2, fas, p53, ced-3; see Lockshin & Zakeri 2001). For example, in the 1990s ced-3 in *C. elegans* was sequenced and it was discovered that it was related to a mammalian protease and that a family of such proteases exist. Known as caspases, these proteases are central components of the apoptosis pathway and their sequence is widely conserved among animals.

The third reason which explains why the field of cell death increased in importance around 1990 was the recognition of the medical importance of apoptosis, which was in turn based on the discovery that apoptosis played a crucial role in several organismic functions. Central for this realisation were the discoveries that apoptosis is very common, is genetically based, is intimately related to the immune system and is important in cancer research. For example, it has been discovered that it is regulated by the p53 tumour suppressor gene. This established a relation between apoptosis and 'differentiation and maintenance of the immune system' (Lockshin & Zakeri 2001, 549). Thus, in the last decades the clinical relevance of apoptosis became evident.

3.2.7   Biochemical Pathways of Apoptosis

The causal pathway of apoptosis can nowadays be characterised not only morphologically, but also biochemically. To illustrate the kind of understanding of the mechanism of apoptosis that we now possess and to compare it with the cytological description outlined earlier, let us briefly

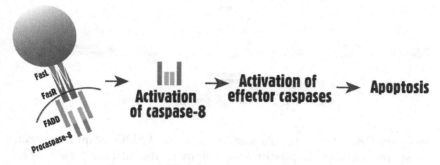

Figure 3.2 The extrinsic pathway of apoptosis.

review the description of the mechanism at the biochemical level, focusing on the mammalian apoptosis pathway (for a more detailed description, see, e.g., Shiozaki & Shi 2004; Cairrão & Domingos 2010).

A central part of the pathway is a process called the 'caspase cascade', which is a positive feedback cascade that involves the activation of caspases, a type of enzyme that when activated performs proteolysis. Caspases exist in the cytoplasm under normal conditions, but not in an active form. Apoptosis occurs when some caspases are activated. Active caspases in turn activate more caspases that eventually break down the cell, producing the morphological changes of apoptosis mentioned earlier.[3] There are two distinct signalling apoptotic pathways: the intrinsic pathway, where the initial apoptotic signal that ultimately leads to the activation of the caspase cascade comes from inside the cell, and the extrinsic pathway, where the initial signal comes from outside.

Here is a brief description of the biochemical mechanism underlying the extrinsic pathway of apoptosis (see Figure 3.2). The initial signal of the extrinsic pathway is the binding of an extracellular ligand to a death receptor, which is an intermembrane protein that has a domain within the cell. For example, T-lymphocytes have a Fas ligand that can bind to the Fas receptor, a protein located on the surface of the cell. The binding of, for example, the Fas ligand to the Fas receptor is the signal for the cell to commit suicide. The Fas receptor has a domain within the cell (Fas-associated death domain [FADD]); when the ligand binds to the Fas

---

[3] Apoptosis can also be produced without a caspase cascade. The central component of this pathway is the apoptosis-inducing factor (AIF), which is located in the intermembrane space in mitochondria. If the cell is damaged, AIF is released into the cytoplasm and moves into the nucleus; it then binds to DNA and destroys it (e.g., neurons can commit apoptosis via this process).

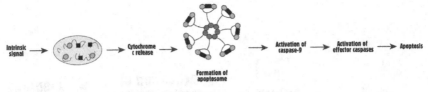

Figure 3.3   The intrinsic pathway of apoptosis.

receptor, FADD is activated and recruits the FADD adaptor protein. Next, procaspase-8 or procaspase-10 binds to the adaptor protein. The formation of this complex, known as the death-inducing signalling complex (DISC), is the signal for the caspase cascade. The procaspases are cleaved and form active caspases 8 and 10. These are the initiator caspases that lead to the activation of effector caspase-3, and so the positive feedback loop of the caspase cascade is activated, which breaks down intracellular proteins.

The biochemical mechanism underlying the intrinsic pathway of apoptosis can be summarised as follows (see Figure 3.3). A central component of the intrinsic pathway is the bcl-2 family of proteins, which include pro-apoptotic and anti-apoptotic proteins. Normally, in the cell there is equilibrium between pro-apoptotic bcl-2 proteins and anti-apoptotic ones, and apoptosis via the intrinsic pathway is prevented. This equilibrium can be disrupted when, for example, the DNA of the cell is irreparably damaged, or the cell stops receiving survival signals. When this happens, other pro-apoptotic bcl-2 proteins are synthesised (so-called BH3-only proteins), which bind to anti-apoptotic bcl-2 proteins. But then, the pro-apoptotic proteins that normally exist within the cell are not inhibited by the anti-apoptotic ones, and can activate the caspase cascade, by causing a central event in the intrinsic pathway, that is, the release of cytochrome c, a protein that normally exists in the intermembrane space of mitochondria, into the cytosol.

This release happens when, because of the disruption of the equilibrium between pro-apoptotic and anti-apoptotic proteins, pro-apoptotic bcl-2 proteins (Bax, Bak) aggregate and form channels in the outer mitochondrial membrane. In normal circumstances where there is a balance between anti-apoptotic and pro-apoptotic proteins, anti-apoptotic proteins (like Bcl-2 and Bcl-xL) bind to the pro-apoptotic ones, thereby stopping them from forming the channels. But when, for example, DNA is damaged and BH3-only proteins are synthesised, they bind to the anti-apoptotic ones, thereby inhibiting the inhibitors and so causing the channels to form. This in turn leads to the release of cytochrome c into the cytoplasm, where it binds to apoptotic protease factor 1 (Apaf-1), which causes Apaf-1 proteins to form a complex called the apoptosome. This activates initiator caspase-9, which then activates effector caspase-3, thereby activating the positive feedback loop of the caspase cascade that breaks down intracellular proteins.[4]

These biochemical descriptions of the signalling apoptotic pathways illustrate the point that there may exist different theoretical descriptions of one and the same mechanism. But another important point here is that the description at the biochemical level is richer than the cytological one in a crucial sense: it provides specific causal information; that is, it identifies difference-making relations among various components of the pathway. Knowing these difference-making relations is important for discovering types of interventions that can be made in order to control the outcome of the process. It is easy to see that the more detailed the description of the causal pathway, the more options for such interventions one has. So, knowing the biochemical description of the apoptosis pathway is important for discovering ways one can intervene to induce apoptosis for therapeutic purposes (for the relations between apoptosis and cancer treatment, see, e.g., Wong 2011). This focus on the potential intervention on causal pathways that is central in biomedicine, then, leads to an important desideratum that theoretical descriptions of causal pathways should have: they should be such so as to provide specific causal information that makes interventions possible (for an account of mechanism in medicine along similar lines, see also Gillies 2017, 2019).

3.2.8   Causal Mechanism as a Generalised Pathway Concept

Thus far we have examined the identification of the mechanism of apoptosis, and, based on this example, we have claimed that mechanisms in biology are causal pathways that are theoretically described.[5] There is

---

[4] Of course, this is a simplified version of the molecular pathways that are nowadays known, which are much more complicated. For example, we now know about the existence of inhibitors of apoptosis proteins (IAPs) that inhibit the activity of caspases by binding to them (via a zinc binding domain). Other mitochondrial proteins, however, can inhibit IAPs (e.g., SMAC, DIABLO) by binding to them when they are released from mitochondria (in case of mitochondrial injury). Other mitochondrial proteins (e.g., Htra2/Omi, AIF, endonuclease G) can cleave IAPs when released (cf. Wong 2011).

[5] Again, this should not be taken to mean that mechanisms are some kind of theoretical description; mechanisms for us are causal pathways and it is possible for the same causal pathway to have multiple descriptions. We add that mechanisms are theoretically described so as to highlight that they are not described in metaphysically loaded terms, as in the standard general characterisations that new mechanists offer.

another route to arrive at this notion, this time using the biochemical descriptions of the mechanism of apoptosis discussed above.

The intrinsic and extrinsic signalling pathways of apoptosis are examples of *biological pathways*. Biological pathways are central in molecular biology. According to the National Human Genome Research Institute, 'a biological pathway is a series of actions among molecules in a cell that leads to a certain product or a change in the cell' (www.genome.gov/about-geno mics/fact-sheets/Biological-Pathways-Fact-Sheet). Biological pathways are, for example, involved in metabolism (metabolic pathways), in the regulation of genes (gene-regulation pathways) and in transmitting signals (signal transduction pathways).

Metabolic pathways synthesise various molecules by utilising energy or release energy by breaking down complex molecules. Glycolysis, the pathway that converts glucose into pyruvate, is a central example. Another simpler example is the pathway in *Escherichia coli* that synthesises the amino acid isoleucine. This pathway has five steps that are catalysed by enzymes and, as is typical in biological pathways, exhibits feedback inhibition: if the amount of isoleucine produced by the cell is more than is needed, the accumulated isoleucine inhibits the activity of the enzyme that catalyses the first step in the pathway, and so the production of isoleucine decreases. Metabolic (and other) pathways are of course not discrete, but interact in complex ways with one another. As Nelson et al. put it,

> although the concept of discrete pathways is an important tool for organizing our understanding of metabolism, it is an oversimplification.... Metabolism would be better represented as a meshwork of interconnected and interdependent pathways. A change in the concentration of any one metabolite would have an impact on other pathways, starting a ripple effect that would influence the flow of materials through other sectors of the cellular economy. (2008, 28)

Signal transduction pathways enable cells to perceive and respond to changes in their environment. In signalling pathways, cells receive signals from the exterior environment or from the interior of the cell (e.g., when DNA is damaged), which lead to a series of chemical reactions that ultimately produce a cellular response to the initial stimulus. The intrinsic and extrinsic pathways of apoptosis we have seen are examples of signalling pathways. Last, gene-regulation pathways involve the regulation, that is, the turning on and off, of genes and can form complex gene-regulatory networks. These networks govern expression levels of mRNA and proteins and are central in ontogeny.

To illustrate, let us consider a part of the p53 signalling pathway. This example also shows the interactions between various pathways, as it is one of the ways the intrinsic apoptotic pathway can be initiated. When DNA is damaged, proteins (e.g., ATM, ATR) that can sense DNA damage become activated. These proteins are serine/threonine kinases; that is, they add phosphate groups to serine and threonine residues of other proteins. In particular, they phosphorylate two other serine/threonine kinases (ChK1 and ChK2), which become active and which in turn phosphorylate p53. Also known as the 'guardian of the genome', p53 is a tumour suppressor protein that is continuously produced within cells. In normal circumstances, p53 is prevented from acting: MDM2 binds to p53, deactivating it and also targeting it for ubiquitination; a ubiquitin group binds to the MDM2-p53 complex, and the complex is transferred to the proteasome where it is destroyed. However, when p53 is phosphorylated, MDM2 cannot bind and so p53 can form tetramers and can act as a transcription factor increasing the expression of certain genes, such as genes that code for proteins involved in DNA repair. Also, it can lead to the expression of p21, a tumour suppressor protein that arrests the whole cell cycle (thus preventing mitosis). If the levels of p53 remain high, this means that the cell cannot repair the damage; the only solution is then for the cell to commit suicide. So, p53 triggers apoptosis via the expression of pro-apoptotic proteins that cause the activation of the intrinsic signalling pathway of apoptosis.

This suffices to illustrate the central importance of biological pathways in all biological functions. The question that we want to pose now is the following: Are biological pathways mechanisms? We think that it is clear from how biologists use the notion that pathways are taken to be kinds of mechanisms. Let us illustrate this claim with a couple of examples from various biological fields.

A first example is, again, apoptosis: the pathways of apoptosis are also referred to as 'mechanisms' by biologists. As a second example, consider the following passage from Rang et al.'s pharmacology textbook:

> Cyclic AMP regulates many aspects of cellular function including, for example, enzymes involved in energy metabolism, cell division and cell differentiation, ion transport, ion channels and the contractile proteins in smooth muscle. These varied effects are, however, all brought about by a *common mechanism*, namely the activation of protein kinases by cAMP – primarily protein kinase A (PKA) in eukaryotic cells. Protein kinases regulate the function of many different cellular proteins by controlling protein phosphorylation (2016, 33; emphasis added).

Here, the activation of protein kinases by cAMP, characterised as a mechanism, is part of the cAMP-dependent pathway, which is a signalling pathway. Further examples are phrases such as 'metabolic mechanisms' and 'transduction mechanisms' to refer to metabolic and signal transduction pathways, respectively.

Here is another example, this time from developmental biology. According to Slack et al. (2005, 132),

> Vertebrate embryos are more or less bilaterally symmetrical, and so for many years the nature of the mechanism producing asymmetry was mysterious. In 1995 it was shown that some key genes (NODAL, sonic hedgehog (SHH), and activin receptor IIa (ACVR2A)) have asymmetrical expression patterns in the early chick, and regulated each other to form a pathway linking an initial symmetry-breaking event to the final morphological asymmetry of the heart and viscera.

Here, the mechanism that produces left-right asymmetry is said to be a pathway linking an initial symmetry-breaking event to the final outcome that is to be explained, that is, morphological asymmetry.

We think that these examples suffice to show that pathways are treated in biology as kinds of mechanisms. But does the converse also hold; that is, are all mechanisms in biology biological pathways? It may be argued that the notion of mechanism in biology is broader than the notion of a biological pathway. Thus, consider again the previous definition of a biological pathway, according to which a pathway is defined as 'a series of actions among molecules in a cell that leads to a certain product or a change in the cell'. Such a definition may imply that a biological pathway is a specifically *biochemical* concept. In contrast, the notion of mechanism can be viewed as a more general notion. For example, when Alberts et al. write about the 'fundamental mechanisms of life – for example, how cells replicate their DNA, or how they decode the instructions represented in the DNA to direct the synthesis of specific proteins' (2014, 22); the 'universal genetic mechanisms of animal development' (p. 39); or the 'fundamental mechanisms of cell growth and division, cell–cell signaling, cell memory, cell adhesion, and cell movement' (p. 1154) – it may be argued that it is not a specifically biochemical conception they have in mind. In addition, if a pathway is viewed as a biochemical concept, then physiological or pathological mechanisms cannot be biological pathways. Last, if a biological pathway only involves a series of changes within a cell, then mechanisms that involve many cells (e.g., morphogenetic mechanisms) cannot be identical to pathways.[6]

Even if this is the case and the notions of a biological pathway and a biological mechanism are not identical, we think that the biochemical notion can be straightforwardly generalised to arrive at a general notion of a causal pathway that, as we are going to argue throughout the book, is present in biology (and elsewhere). According to this more general notion, which we can call the 'generalised pathway concept', a mechanism is a sequence of causal steps (and not just of actions among molecules) that lead from an initial stimulus or cause to an effect (and not just to a cellular response) and is not confined within a cell, but can involve many cells, and more generally can include components from all levels of biological organisation.[7] Such pathways, then, can be described in molecular, cytological or other (even non-biological, e.g., psychological or sociological) terms. It is in this more generalised sense that the cytological descriptions of apoptosis by Kerr et al. are causal pathways. Some justification for the presence of a generalised pathway concept is the fact that pathway-talk is used in biology even in areas outside molecular biology. For example, biologists talk about 'cytological pathways', 'ecological pathways' and 'evolutionary pathways'. Our claim, then, which will be further corroborated in subsequent chapters that will discuss various other examples of mechanisms, is that mechanisms in biology can be identified with causal pathways in this more general sense.[8] According to this generalised pathway concept, then, or *Causal Mechanism*, as we prefer to call our account:

(CM) a mechanism = a theoretically described causal pathway.

[6] Such considerations may underlie the claim made by Giovanni Boniolo and Raffaella Campaner (2018) that mechanisms and pathways in molecular biology are not identical notions. See also Ross (2021) for a treatment of the notion of pathway in biology that shares some similarities with our account. Ross claims that pathways and mechanisms are different notions. In part, she arrives at this conclusion because she uses a notion of mechanism with which we do not agree. She writes: 'When [biologists] use the mechanism concept they often suggest that some biological phenomenon can be understood as a kind of machine or mechanical system.... This machine analogy encourages thinking of biological phenomena as having component parts that are spatially organized and that causally interact to produce some behavior of the system' (134). While we tend to agree that this is how many new mechanists think about mechanisms, we do not think that this is the best way to characterise the notion of mechanism as used by biologists. Ross also suggests that pathways abstract from causal detail, while mechanisms do not. We disagree, as many mechanisms abstract from causal detail too. Our view is that no clear distinction is to be found in biology between mechanism-talk and pathway-talk. Notions such as mechanisms, pathways, cascades and mediators all are instances of the same general concept of causal pathway.

[7] We discuss the important issue of the relationship between causal pathways and levels of organisation in detail in Chapter 9.

[8] Although in biology mechanisms as causal pathways often have function, sometimes they do not. See Section 3.4 and Chapter 4, Section 4.5.2.2, for details.

The advantage of this strategy to arrive at CM is that we start from a concept that is clearly central in biology and for which there even exists a definition, that is, the notion of a biological pathway, and we generalise it, in order to arrive at the more general notion of CM. We then use CM to illuminate the concept of a biological mechanism, for which no explicit definition exists in the biological literature (which, incidentally, suggests that mechanisms and biological pathways in the narrow sense are not viewed as identical concepts by biologists). We can thus illuminate the more general concept (i.e., the concept of a mechanism), which we take to be a concept-in-use central in biological practice, without using ordinary notions of what a mechanism is (such as the notion of a 'machine' as used in everyday contexts) or introducing metaphysical categories that seem to be extraneous to biological practice, but by using concepts already present in biology (i.e., the biochemical concept of a pathway).

Let us now turn to some objections to CM, focusing again on the example of apoptosis. It might by objected that the features of mechanisms in biology identified in Section 3.2.5 – that is, that mechanisms are causal pathways theoretically described – are not in fact exemplified in the case of the signalling pathways of apoptosis, as well as in other cases of biological pathways. That is, it might be claimed that apoptosis (1) is not linear, (2) has no obvious start and end points, (3) involves homeostasis and (4) is such that spatiotemporal organisation is crucial. Collectively, it might be argued, these features cannot be accounted for without a more metaphysically inflated account of mechanisms, such as those introduced in Chapter 2.

By way of reply, it should be noted that all the above features are compatible with our account of mechanisms as theoretical descriptions of causal pathways. First, pathways are processes; but it does not follow that they have to be 'linear' processes. Apoptosis as characterised cytologically might seem 'linear', that is, as a stepwise process with no negative or positive feedback loops. However, its biochemical description is far more complex, with multiple pathways and complex causal cycles, as is typically the case in biological pathways. But in any case, we still have, ultimately, a web of sequences of causal steps related by difference-making relations. Second, we do not take it as a general requirement that a process has to have well-defined starting and end points; this is something to be determined by biological practice. In the case of the extrinsic pathway of apoptosis, for example, the starting point of the process is specified by the binding of a ligand, such as the Fas ligand, to a death receptor. Third, homeostasis, which itself involves negative feedback loops, can also be described in terms of difference-making relations among the components of the pathways underlying the homeostatic process.[9] Fourth, the spatio-temporal organisation is indeed crucial for the functioning of any causal pathway, and because of this it is typically included in the theoretical description of the pathway: a detailed theoretical description of the causal pathway will state the spatiotemporal relations among the various components of the pathway that are required for the proper functioning of the mechanism.

To sum up, the example of apoptosis shows that some salient features of mechanisms in biology are the following: mechanisms are causal pathways described in theoretical language that have certain functions; these descriptions can be enriched by offering more detailed or fine-grained descriptions; the same mechanism can then be described at various levels using different theoretical vocabularies (e.g., cytological vs biochemical descriptions in the case of apoptosis);[10] last, the descriptions of biomedically important mechanisms are often such that they contain specific causal information that can be used to make interventions for therapeutic purposes.

3.3   Mechanisms of Cell Death

An important issue that crops up here for CM is how the various causal pathways are identified and distinguished from each other. Could it be the case that some causal pathways are *mechanisms* in a more robust sense while others are *merely* causal pathways? In the case at hand, the issue is how apoptosis can be distinguished from other cell-death processes. As we have noted already, the activation of caspases underlies the morphological changes that occur during apoptosis. However, not all processes of cell death are apoptotic. That is, not all processes that lead to cell death feature the particular sequence of morphological changes associated with apoptosis, nor do all involve caspase activation. This is why in the literature on cell death there are more general terms that are used to refer to the various

---

[9] See Woodward (2002) for a description of the *E. coli* lac operon regulatory mechanism in terms of difference-making; as Woodward notes, difference-making better captures the causal relationships present in negative regulation that is involved in homeostatic mechanisms, which involves cases of so-called double prevention that present difficulties to alternative characterisations of within-mechanism interactions.

[10] It is a widely held view in the philosophical literature on mechanisms that mechanisms form hierarchies (see Povich & Craver 2018 for a discussion of mechanistic levels). While we agree that a mechanism such as apoptosis can be described at various levels, we do not derive any ontological consequences from such talk (see Chapter 9 for our views on levels).

types of processes of cell death: on the one hand, there is 'physiological' or 'regulated' cell death; on the other, 'accidental cell death'. Lockshin and Zakeri explain the difference between the two by noting that when cells die as a result of a process of physiological cell death, 'such deaths are part of the normal function of the organism'; also, in physiological cell death genetic regulation is central. This is not the case in 'necrosis or oncosis which is accidental and in which the cell has no active role' (Lockshin & Zakeri 2001, 545).

Could it then be the case that there is a genuine difference between apoptosis and necrosis such that apoptosis is a mechanism in a more robust sense than just the causal pathway for cell deletion, whereas necrosis is *merely* a causal pathway? To address this issue, let us note first that, as Guido Majno and Isabelle Joris (1995) have pointed out, apoptosis and necrosis should not be juxtaposed: apoptosis is a process that leads to cell death, but necrosis should not be used to refer to such a process. As they stress, there is a distinction between cell death and necrosis: cell death comes about before necrotic changes can be observed. Necrotic changes (e.g., karyolysis, karyorhexis, loss of structure in the cytoplasm) 'are the features of a cell's cadaver, whatever the mechanism of the cell's death, be it ischaemia, heat, toxins, mechanical trauma, or even apoptosis' (Majno & Joris 1995, 11). This has actually been the traditional meaning of necrosis; that is, it refers to changes in tissues that are visible even without a microscope and as such occur after cell death.[11]

This point has been emphasised also by other authors. So, according to Kanduc et al. (2002, 167), it is 'scientifically unjustified' and 'unsound' to compare apoptosis to necrosis, as apoptosis is a *process* that leads to cell death, whereas necrosis refers to changes that occur to cells *after* they die. In general, then, we should distinguish between the processes that a dying cell undergoes (e.g., apoptosis) and the end result of these processes, which is the dead cell. 'Necrosis' should then refer to already dead cells and tissues and the changes that occur *after* cell death (see also Fink & Cookson 2005). This usage of the term 'necrosis' is precisely the one suggested by the Committee on the Nomenclature of Cell Death chartered by the Society of Toxicologic Pathologists to make recommendations 'about the use of the terms "apoptosis" and "necrosis" in toxicity studies' (Levin et al. 1999, 484). In line with the above, this committee recommended that

---

[11] For example, here is how Rudolf Virchow in the nineteenth century describes necrosis in *Cellular Pathology*: 'In necrosis we conceive the mortified [gangrenous] part to be preserved more or less in its external form' (quoted in Majno & Joris 1995, 3–4).

'when dead cells or tissues are observed in a histological lesion, "necrosis" is the appropriate morphological diagnosis, regardless of the pathway by which the cells or tissues died' (p. 486). They conclude: 'This Committee believes that returning to the long-established histopathological standard wherein the word necrosis denotes dead cells in a living tissue (regardless of their phenotype) should help to alleviate the confusion attendant on the notion, held by many, that a dichotomy exists between apoptosis and necrosis' (p. 489). Notably, Robert Sloviter (2002, 22) goes as far as to note that 'the term necrosis is now virtually meaningless because "necrotic" means nothing more than "dead"'.

According to Majno and Joris, 'the major sore spot in the nomenclature of cell death is precisely the lack of a suitable name for cell death that occurs not by apoptosis but by some external agent' (Majno & Joris 1995, 11). And this is precisely the point: cell death might have different causal pathways, and the difference between them is not that one (or some) of them counts as a mechanism while the other does not; rather, the difference is in *how they are described*.

Majno and Joris's own suggestion is to characterise apoptosis in contrast with a specific process of cell death they call 'oncosis' (p. 12). A common non-apoptotic causal pathway that leads to cell death is when groups of cells die of ischaemia (*ischaemic necrosis*). During the causal pathway that leads to ischaemic necrosis the cell swells, and it is in order to capture this swelling process that the authors propose 'oncosis' as a term to refer to this mechanism. This causal pathway can nowadays be characterised in detail: reduced supply of oxygen and nutrients leads to ATP depletion, which ultimately results in protein denaturation, enzymatic digestion due to damaged lysosomes and loss of integrity of the plasma membrane, resulting in influx of water and calcium into the cell, leading to swelling and ultimately rupture of the cell. Also, we know that ischaemia typically activates the causal pathway, but toxic agents can also initiate it. We can then talk about the mechanism of oncosis in the CM sense of the term. Oncosis and apoptosis, then, are two causal pathways (and hence mechanisms) of cell death – by swelling and by shrinkage, respectively.

Can there be a way to distinguish apoptosis from oncosis such that only the former counts as a mechanism? We will consider three distinctions used by researchers of cell death that might be used in order to do this: (1) processes of physiological versus accidental cell death, (2) processes of programmed versus non-programmed cell death and (3) active versus passive processes.

A term widely used to describe a non-apoptotic, non-physiological type of cell death is 'accidental cell death'. By describing a process of cell death as 'accidental', biologists try to capture the idea that, in contrast to apoptosis, this is not a process that occurs under normal conditions, nor does it serve a general homeostatic function within the organism. However, as Majno and Joris (1995) note, apoptosis can also be induced by a variety of 'accidental' causes (e.g., heat, chemical agents, viruses). Levin et al. (1999) also note that 'dead cells having the cytological features of apoptosis can occur in large numbers as a pathological change, e.g., "single cell necrosis" in the liver and lymphocyte necrosis in the thymus, and that these changes can be induced by exogenous events such as exposure to toxicants' (p. 485). So, the very fact that apoptosis can be initiated by 'accidental causes' shows that the right contrast here is not between mechanism and non-mechanism, but rather between physiological cell death and accidental cell death, where 'physiological cell death' refers to a process of cell death that was initiated by physiological stimuli.

What about the programmed/non-programmed distinction? An important point here is that 'programmed cell death' and apoptosis should not be identified; thus, the former cannot be used to distinguish the latter. Programmed cell death is the phenomenon where cells die 'on schedule'; that is, they are programmed to die at a specific time. During development of the chick, for example, the morphology of the wing is produced as a result of the death of groups of cells; there is a 'genetic clock' that determines when the cells will die. But when the time comes for the cell to die, the specific programme that determines the form that cell suicide will take is triggered. The particular form of cell suicide can be apoptosis (indeed, very frequently it is), but it need not be. As Majno and Joris stress: 'The genetic programme of programmed cell death is a clock specifying the time for suicide, whereas the genetic programme of apoptosis specifies the weapons (the means) to produce instant suicide' (1995, 11): the weapon is precisely what we call the causal pathway. Again, the point here is that one cannot use this distinction to distinguish apoptosis as a genuine mechanism.

Biologists, we noted above, have characterised the contrast between processes of cell death along the active/passive lines: apoptosis was described as 'active' from the very beginning of its introduction, and contrasted with the 'passive' necrosis.[12] Perhaps what makes apoptosis a mechanism, then, is precisely that it is an active process. However, what does this distinction really mean? The idea here seems to be that in the case

---

[12] However, see Proskuryakov and Gabai (2010) for the argument that necrosis can in certain cases be considered an active and well-regulated process.

of apoptosis the cell is itself involved in its own demise ('cell suicide'), whereas in the case of oncosis the cell dies as a result of some exogenous influence ('cell murder'). That is, apoptosis involves a 'suicide programme' that is initiated under various circumstances and that is genetically based, in the sense that there exist specific genes that code for various components of the biochemical pathway underlying the apoptotic process (the distinctions regulated vs non-regulated, ordered vs unordered, controlled vs non-controlled appear to be used in a similar way).

To make this clear, Sloviter (2002, 23), after describing the two causal pathways as 'active cell death' (ACD) and 'passive cell death', notes that ACD is active in the sense that it requires 'active intracellular processes for death to result', whereas in passive cell death 'the cell plays no role in its own demise'; that is, cell death is 'immediate and involves no cellular activity', the cause being exogenous to the cell such as 'rapid freezing, aldehyde fixation, heat denaturation, and catastrophic physical destruction'. As such, passive cell death is of little interest since being immediate it 'offers no therapeutic window'.

The important point for us is that this difference between 'active' and 'passive' is merely a difference concerning the details of each causal pathway. Hence, there is no intrinsic difference between the two causal pathways as such: there is nothing particularly active in apoptosis and particularly passive in oncosis. The significant difference from a biological point of view is that because apoptosis involves a 'suicide programme', it can serve a homeostatic function, as argued by Kerr et al. (1972).

To avoid the insinuation that the use of 'active' and 'passive' processes might lead to views about the ontology of causation and mechanisms (e.g., to the distinction between entities and activities, as in Machamer, Darden & Craver 2000), let us see how biologists view the active/passive distinction. Kanduc et al. (2002) say:

> It is frequently assumed that the death of cells can be passive. This non-biological point of view on cell death ignores the role of cell death in cell development and adaptation. It cannot be assumed that 'ordinary' cell death or 'necrosis' is a passive process while the presumed special form of cell death, 'apoptosis' is active. *Both the ante-mortem and postmortem changes are active since both are enzyme-catalysed biochemical reactions.* (pp. 167–8, emphasis added)

So, to call a biological process 'active' should not commit us to a specific metaphysical view about the metaphysics of causation. In particular, it is not a reason to adopt an activities-based understanding of mechanisms as opposed to a difference-making account (see Chapter 6 for a detailed examination of activities as an ontic category).

3.4 Is Mechanism More than the Causal Pathway?

Hence, apoptosis and oncosis can both be considered mechanisms in the CM sense: they are both causal pathways that produce a result (apoptotic and ischaemic necrosis, respectively). However, the history of programmed cell death and apoptosis during the last 60 years might be used to argue that in biological practice what counts as a *biological* mechanism cannot just be a matter of identifying a specific causal pathway. Apoptosis seems to be a special kind of causal process with distinctive features that deserves to be labelled a mechanism. This can even be seen in biologists' description of apoptosis as a 'mechanism of cell death', whereas necrosis or oncosis are not usually described as 'mechanisms'.

Note that the reason apoptosis became a central biological mechanism is not due to some feature internal to the sequence of causal steps that constitutes the apoptotic pathway, but rather due to features that are external to the pathway itself; that is, because it is a key process that controls homeostasis. As we have seen, it is its role within the developing and adult organism, that is, its homeostatic function, that led to the formulation of the concept by Kerr et al in 1972; similarly, it is the discovery of its highly controlled nature and conservation of the genetic sequences of the components of the apoptotic pathway across animals (which shows its central functional importance in animals), as well as the realisation of the close relation between apoptosis and the immune system and cancer that followed the molecular genetic discoveries of the 1980s and 1990s, that gave it the central prominence it deservedly has today as a biological phenomenon.

To put the point differently, what we think the story of apoptosis shows is the following: in the world there are causal pathways for various phenomena; all causal pathways can be deemed mechanisms in the deflationary sense of CM; but not all those causal pathways are *biologically* interesting or significant, even if they occur frequently within organisms. Biologically interesting or significant causal pathways are those pathways that subserve a central *function* within the organism; that is, whether a causal pathway is biologically interesting has to do with features *external* to the pathway itself.

This does not imply that what is biologically interesting is something subjective. Rather, it implies that it is directly related to biological practice: what the community of biologists regards as the basic phenomena that must be explained in order to have both biological understanding and apply our knowledge clinically. Thus, we could say that what makes a causal pathway a specifically biological mechanism is not something internal to the pathway itself; rather, it concerns the functional role of that pathway within the organism. In other words, whether a causal pathway is considered a biological mechanism by biologists has to do with a relational property of the pathway. This relational property concerns how the pathway is related to the other processes that occur within the organism so as to subserve a certain function. In the case of apoptosis, this relational property concerns the homeostatic function that it subserves. This relational property is the difference between a causal pathway like apoptosis and a causal pathway such as oncosis.

However, suppose one were to argue as follows. We should certainly let biological practice itself decide what we should mean by a 'mechanism' in a biological context. If practice has it that a causal pathway is deemed a mechanism by an appeal to external features of the pathway, so be it. CM (the point would be) is false, since

(P-CM) A mechanism is a theoretically described causal pathway +X, where X is some biologically significant external feature of the causal pathway (like its function).

Now, if one were to argue like this, we would not seriously object. We are ready to accept that there may well exist features external to a particular causal pathway and, in particular, features that can be established by looking at biological practice, which determine whether a specific theory–described causal pathway counts as a biological *mechanism*. But, in our view, this attribution of 'mechanism' to certain causal pathways and perhaps not to others would entail that 'mechanism' is an honorific term attached to causal pathways that have certain (external) features. The further scientific question, then, is whether there is evidence that a causal pathway is a 'mechanism' in this sense or not.

In our view, the choice between CM and P-CM is not particularly significant: to adopt P-CM is to claim that we allow a distinction between a causal pathway for a phenomenon P and a specifically *biological* mechanism, where the difference between the two concerns an external feature of the respective causal pathways. Be that as it may, the important point is that both views are licensed by Methodological Mechanism: to be committed to either option, one need not be committed to some metaphysical view about causation or the ontology of mechanisms. As we will argue in Chapter 4, there is no need to do this in order to understand scientific practice.

Could someone insist that there is some other feature that differentiates a causal pathway from a mechanism? A possibility here is to adopt the requirement of causal modularity (cf. Woodward 2002). Causal modularity may be seen as the criterion that determines whether a process counts as machine-like or not; so, perhaps 'mechanism' should be used only for causal processes that exhibit causal modularity. While modularity can be important in many cases as a requirement for a causal representation of a system, the major disadvantage of this view is that many instances of 'mechanisms' in biology turn out not to be such, since they are not modular; apoptosis is a case in point (cf. Cassini 2016). So, adopting this view necessitates abandoning taking scientists' talk of mechanisms at face value.

Could someone insist that there is some other feature that differentiates a causal pathway from a mechanism? A possibility here is to adopt the requirement of causal modularity (see Woodward 2002). Causal modularity may be seen as the criterion that determines whether a process counts as machine-like or not; so, perhaps 'mechanism' should be used only for causal processes that exhibit causal modularity. While modularity can be important in many cases as a requirement for a causal representation of a system, the major disadvantage of this view is that many instances of 'mechanisms' in biology turn out not to be such, since they are not modular; apoptosis is a case in point (see Casini 2016). So, adopting this view necessitates abandoning taking scientists' talk of mechanisms at face value.

3.5   What Does the Case of Apoptosis Show?

What can we learn from the case of apoptosis? In our examination of this case we have use two strategies. The first strategy was to examine the cytological description of the mechanism of apoptosis in order to extract some main features of mechanisms in biology. The most important such feature that we found is that mechanisms are processes, that is, sequences of causal steps that we called causal pathways. The second strategy was to examine the biochemical pathway concept and to generalise it, in order to arrive at a generalised pathway concept. This strategy enabled us to argue that all causal pathways are mechanisms and all mechanisms are causal pathways. Both strategies lead to a general notion of a causal pathway (or mechanism) that we think is the typical notion of mechanism in the life sciences and elsewhere.

But of course apoptosis, it can be argued, is just one example. A full defence of the claim that mechanisms are identical to causal pathways would require the examination of many more cases. In a sense, this is correct. In the subsequent chapters of the book, we will examine various other cases of mechanisms and argue that they just are causal pathways. By the end of the book, our thesis that mechanisms are causal pathways will have become much more convincing.

However, the example of apoptosis is not just one case study among others. We take this case to be paradigmatic. First, the pathways of apoptosis illustrate a central concept in molecular biology, the concept of a molecular pathway. There are many specific examples of molecular pathways in current biology that are similar to the molecular pathways of apoptosis, some of them mentioned in Section 3.2.8. The second strategy, in particular, involved not just the particular case of the signalling pathways of apoptosis, but a general biochemical pathway concept that is prevalent in molecular biology. It is this general molecular concept that we generalised further in order to arrive at our generalised pathway concept and finally argue that all pathways are mechanisms, and vice versa. Nothing in this strategy depends on the particular case of molecular pathway chosen.

A second reason why we take apoptosis to be paradigmatic is that (as has been mentioned in Section 3.2.1) it has some very important features, compared with other cases. Apoptosis nicely illustrates how a *new* mechanism was first introduced, and thus it clearly shows what is involved in first identifying and describing a biological mechanism. Moreover, it illustrates several other features of mechanisms, such as that they can be described at many different levels, their relation with functions and how they are differentiated from similar mechanisms (such as necrosis). Being a mechanism that is central in various biological fields, apoptosis also illustrates a concept of mechanism that is common across biology. Last, it is a fundamental biological mechanism, but relatively unknown among new mechanists. In general, then, the example of apoptosis is for us a lot more than a case study.

It is also important to note that the case of apoptosis, as well as other cases of pathways and mechanisms described in this chapter and throughout the book, are not intended as premises of an inductive argument, the conclusion of which is that mechanisms are to be equated with causal pathways. If our argument were an inductive one, a lot more examples would be required to establish its conclusion. However, we do not view the generalisation from apoptosis to CM as a general concept of mechanism that applies widely within biology, as an induction; we take the apoptosis case as a paradigmatic illustration rather than as the starting point of an inductive argument. So, we think that the important issue is not just to include more cases but to include important and paradigmatic cases that illustrate our view.

We think, then, that the discussion in this chapter shows that the generalised pathway concept is prevalent in biology (subsequent chapters will further corroborate this claim). What remains to be shown is whether this concept suffices in order to capture all typical uses of 'mechanism' in biology. In other words, are there mechanisms that are not just causal pathways (nor pathways with a certain function), but something more? To answer this question, more cases should be examined. It is also important to discuss cases that are disciplinarily distinct, in order to be able to argue that mechanisms as causal pathways are to be found across the life sciences and elsewhere. In the subsequent chapters we will have the opportunity to apply the generalised pathway concept to various cases. For example, we think that this concept is easily applied to mechanisms in evolutionary biology, which have been thought to be problematic cases for other general characterisations of mechanism such as the MDC account (see Skipper & Millstein 2005; we will come back to this issue in Chapter 10).

Last, does the analysis in this chapter provide reasons to prefer CM to other accounts of mechanism? One could argue that if Minimal Mechanism or the MDC account can capture this case equally well, then the example of apoptosis does not help us very much. Let us first note that the main purpose of this chapter is not to claim that CM captures the case of apoptosis better than other accounts of mechanism. Our main aims here, and the reasons why we discuss at such length the apoptosis case, are, first, to examine the notion of a biological pathway – a systematic examination of this notion is absent from the literature and part of our aim in this chapter is to provide such an analysis – and, second, to use it in order to examine the concept of a biological mechanism. We will further develop CM and provide some reasons to prefer it to metaphysically inflationary accounts in Chapter 4. In Chapters 5 and 6 we will criticise inflationary accounts in detail and in Chapters 8 and 9 we will provide reasons why CM is preferable to Craver's account. So, we have various other reasons to prefer CM to other accounts of mechanism that are only indirectly related to our analysis of the apoptosis case. However, let us close this chapter by briefly mentioning two main reasons why the analysis in this chapter can be used to single out CM (we will come back to these points in Chapter 4). First, the examples of mechanisms examined in this chapter strongly suggest that typical and paradigmatic mechanisms in biology are causal pathways, as opposed to complex systems (Glennan 1996, 2002) or Craver's (2007a) 'constitutive' mechanisms. Second, while accounts that put more emphasis on the processual nature of mechanisms such as the

MDC account are consistent with the apoptosis case, if CM is also consistent with this example and CM is the more minimal view, this surely must be a reason to prefer CM over ontologically more inflated accounts. The important issue here is to identify the features of the concept-in-use. In particular, we take our case study to show that, as will become clearer in Chapter 4, the excess ontological content of the MDC and other accounts does not play a role *within scientific practice*.

CHAPTER 4

Mechanisms as Causal Pathways

4.1 Preliminaries

Our key thesis, which we call *Causal Mechanism*, is that a mechanism is a causal pathway that is described in theoretical language, where the pathway is underpinned by networks of difference-making relations. In advancing CM, we will show:

1. that there is indeed a common notion of mechanism in virtue of which different kinds of mechanisms (at least in life sciences) count as such;
2. that this general account is ontologically minimal (in particular, it incorporates only a notion of causation as difference-making); and
3. that non-causal constitutive relations are not required to understand what a mechanism is in scientific practice.

Since we are offering a causal account of mechanism, we take it that the concept of causation is central in characterising a mechanism. We argue, however, that causation is a relation of robust dependence among particulars and that stronger accounts of causation are not required to understand the concept of a causal pathway.

4.2 Causal Mechanism: Three Theses

According to current mechanistic approaches to causation, causes *produce* their effects, where production is cashed out in terms of mechanisms. Mechanisms are taken to be complex systems or, in general, acting entities, which are composed of parts and have internal structure or organisation and certain spatiotemporal locations.[1] The mechanism has a characteristic

---

[1] The term 'complex system' is used by Glennan (1996, 2002), but then he drops it. MDC (2000) and Craver (2007a) give more processual accounts. See also Krickel (2018, chapter 2), who distinguishes between the 'Complex System Approach' (Bechtel & Abrahamsen, early Glennan) and the 'Acting

Figure 4.1   A mechanism as causal pathway.

behaviour in virtue of the properties, dispositions or capacities of its parts as well as in virtue of how these parts are organised and interact with each other. *What* the mechanism is doing (its characteristic activity, its behaviour or its output) is caused and explained by the details of *how* it is doing it. These details include the internal workings of the mechanism.

We shall critically discuss the mechanistic approach in some detail in Chapter 5. Suffices it to say that CM is not a version of a mechanistic account of causation; rather, it takes causation (as difference-making) to be conceptually prior to the notion of a mechanism (as a causal pathway). Let us now examine the main commitments of CM in more detail.

We have used the case of apoptosis as a benchmark to develop a characterisation of the concept of mechanism as used in biological practice. The core idea of this practice-based account is that mechanisms in biology are equivalent to stable causal pathways, described in the language of theory, which bring about a certain effect and (at least in some cases) perform a certain function (see Figure 4.1). We have called this account of mechanism *Causal Mechanism*:

(CM) A mechanism is a theoretically described causal pathway.

By way of further clarifying CM, it is useful to distinguish between three theses that together constitute CM. The core thesis of CM is that mechanisms are causal pathways. But, moreover, for CM causal relations among the components of the pathway are to be viewed in terms of difference-making. We will call this thesis the *difference-making* thesis of CM. In addition, according to CM the causal pathway has to be described in theoretical terms and not in terms of one's favourite metaphysics. CM then remains agnostic about the underlying metaphysics of mechanisms. We will call this thesis the *metaphysical agnosticism* thesis of CM. Let us examine these three theses in more detail.

The mechanisms-as-pathways thesis differentiates CM from most accounts of mechanism found in the literature. As already noted, some

Entities Approach' (Machamer et al.). In saying that mechanisms in biology are to be viewed as causal pathways, we disagree with both of these approaches.

new mechanists take mechanisms to be complex systems. Even if this is not explicitly incorporated in a general characterisation of mechanism, mechanisms are often viewed as kinds of organised entities or structures. The underlying intuition here seems to be the everyday notion of a mechanism or a machine: an organised structure made of parts that interact. But if the aim is to identify the notion of mechanism implicit in biology, we think that this is misleading, since mechanisms and systems, for biologists, are different things: mechanisms, we think, are best viewed as causal pathways, and it is strange to call a causal pathway a 'system'. In contrast to the view of mechanisms as complex systems, then, CM takes mechanisms to be processes. The main reason for viewing mechanisms in biology in terms of processes rather than in terms of systems, entities or machine-like structures, then, is that this is the notion that is at work in central examples of biological practice. The case of apoptosis examined in Chapter 3 provided support for this view.

It is important also to note, in relation to this core thesis of CM, that although we call our position *Causal Mechanism* in order to stress the close relationship between mechanism and causation, as well as the fact that causation as a concept is prior to mechanism, on our view a mechanism just is a causal pathway and every causal pathway is a mechanism. In other words, CM *identifies* mechanisms with causal pathways, both ontologically and conceptually; mechanisms and pathways are not distinct notions. Hence, it is not the case that some mechanisms are *causal* whereas others are not.

The view that comes closer to CM regarding the core thesis of mechanisms as causal pathways is the recent account of Donald Gillies about mechanisms in medicine. He takes mechanisms in medicine to be causal pathways connecting a cause with a particular effect. Gillies sums up his account as follows: 'Basic mechanisms in medicine are defined as finite linear sequences of causes ($C_1 \rightarrow C_2 \rightarrow C_3 \rightarrow \ldots \rightarrow C_n$), which describe biochemical/physiological processes in the body. This definition corresponds closely to the term "pathway" often used by medical researchers. Such basic mechanisms can be fitted together to produce more complicated mechanisms which are represented by networks' (Gillies 2017, 633).

In discussing the biochemical descriptions of the signalling pathways of apoptosis in Chapter 3, we have claimed that causal relations within a pathway are best understood in terms of difference-making. This is then a basic tenet of CM, which can now be expanded as follows: mechanisms are causal pathways involving difference-making relations among the components of the pathway. The *difference-making thesis* of CM differentiates

CM from accounts of mechanism that characterise within-mechanism interactions in terms of activities or the manifestation of powers.

This second thesis of CM is shared by other accounts of mechanism. For Woodward (2002), difference-making is required to account for within-mechanism interactions. As he puts it, 'components of mechanisms should behave in accord with regularities that are invariant under interventions and support counterfactuals about what would happen in hypothetical experiments' (Woodward 2002, 374). In his 2002 paper, Glennan has adopted this view. Peter Menzies (2012) has used this interventionist approach to causation to give an account of the causal structure of mechanisms. Gillies also views causation in terms of difference-making. From those accounts, Gillies's view more closely resembles CM.[2] For Glennan, in particular, it is not the case that causation is conceptually prior to mechanism, since he uses mechanisms to offer a theory of causation, as we will see in more detail in Chapter 6.

The thesis of *metaphysical agnosticism* is a novel feature of CM that differentiates it from all extant accounts. Because of this thesis, the description of the pathway, according to CM, has to be given in the theoretical terms of the specific scientific field (or fields) and not in terms, for example, of entities and activities. Moreover, it is because of this thesis that CM puts the methodological role of mechanism at the centre; this will lead to the general framework that we will call Methodological Mechanism, which will be developed in Chapter 10.

It is important to stress that the theses of *metaphysical agnosticism* and *difference-making* are logically independent, as one can hold one, but not necessarily the other. For example, one can understand within-mechanism causal relations in terms of difference-making, but at the same time use metaphysical categories to understand what a mechanism is, or subscribe to a specific view about the truth-makers of causal relations. For example, one can hold a neo-Aristotelian view according to which causal relations supervene on the powers of entities, and provide an analysis of mechanisms in neo-Aristotelian terms. In advancing CM, we will reject this option. We

---

[2] Woodward's (2011) account of mechanisms also shares similarities with CM. Woodward is skeptical that there exists a common notion of mechanism present across diverse scientific fields. He writes: 'I believe it is a mistake to look for an account of "mechanism" that will cover all possible interpretations or applications of this notion. A better strategy is to try to elucidate some core elements in some applications of the concept and to understand better why and in what way those elements matter for the goals of theory construction and explanation' (p. 410). We think that mechanisms as causal pathways capture a core notion of mechanism present in many different scientific contexts.

will argue that in giving a characterisation of mechanism as a concept-in-use, one need not be committed to a specific view concerning the metaphysics of mechanisms: mechanism in our sense is a concept used in scientific practice and as such it is primarily a methodological concept. This will pose a challenge to those philosophers who adopt metaphysically inflationary accounts of mechanism: given the adequacy of CM to capture the concept-in-use, the burden is on the defender of a particular metaphysical characterisation of mechanism to say why a methodological account such as CM is not enough and why it should be inflated with metaphysical categories such as entities and activities – we will come back to this point in Section 4.5.

Conversely, one may accept *metaphysical agnosticism* by remaining neutral on the underlying metaphysics, without subscribing to a difference-making account of within-mechanism interactions. This shows that the thesis of *difference-making* needs independent justification. Our examination of the discovery of the mechanism of scurvy in Section 4.3 will start to provide such a justification and will show that *difference-making* is central for understanding mechanism as a concept-in-use.

It is also possible to hold *metaphysical agnosticism* (and perhaps also *difference-making*) and reject the core thesis of CM that mechanisms are causal pathways, in favour of the view that mechanisms are complex entities or, more generally, kinds of organised entities. That is, one could remain neutral on how to understand entities, activities and organisation from a metaphysical point of view. An important point here is that commitment to the minimal characterisation of organised entities and activities is more natural if one holds the complex systems view of mechanisms and not the view that mechanisms are causal processes. But, moreover, 'activities', for example, are typically introduced as a novel ontological category; hence, the elements of the minimal characterisation are too metaphysically loaded and very hard to read in such a neutral way. In Section 4.4, where we focus on the inflationary accounts of mechanisms, we will examine this important point in more detail. First, however, let us examine a second case study that will serve to further illuminate the view of mechanisms as causal pathways and the thesis of *difference-making*, that is, the case of scurvy.

4.3 The Case of Scurvy

In this section we look in some detail at the history of the discovery of the mechanism of scurvy. We use this case to further illustrate CM and, in particular, the view that difference-making suffices to capture causal relations among the components of the pathway. The key point will be that difference-making is what is important in biological practice when a new pathway is identified. Moreover, we will use the example of the discovery of the mechanism of scurvy in order to argue that mechanistic evidence in science is evidence about difference-making relations.[3]

4.3.1 The Discovery of the Mechanism of Scurvy

Scurvy, we now know, is a disease resulting from a lack of vitamin C (ascorbic acid). The absence of vitamin C in an organism causes scurvy, which starts with relatively mild symptoms (weakness, feeling tired, sore arms and legs) and if it remains untreated it may lead to death. If we take seriously the thought that absences, qua causes, are counterexamples to mechanistic causation, we should conclude that there is no mechanistic explanation of scurvy. But this would be clearly wrong. What is correct to say is that the lack of vitamin C disrupts various biosynthetic causal pathways, that is, mechanisms, for example, the synthesis of collagen. In the latter process, ascorbic acid is required as a cofactor for two enzymes (prolyl hydroxylase and lysyl hydroxylase) that are responsible for the hydroxylation of collagen. Some tissues such as skin, gums, and bones contain a greater concentration of collagen and thus are more susceptible to deficiencies. But ascorbic acid is also required in the enzymatic synthesis of dopamine, epinephrine and carnitine. Now, humans are unable to synthesise ascorbic acid, the reason being that humans possess only three of the four enzymes needed to synthesise it (the fourth enzyme seems to be defective). Hence humans have to take vitamin C through their diet.[4]

The disrupted causal pathways that prevent scurvy can be easily accommodated within the difference-making account of causation. Had vitamin C been present in the organism X, X wouldn't have developed scurvy. In fact, the very causal pathway can be seen as a network of relations of dependence (or difference-making). Abstractly put, had vitamin C been present in human organism X, X's lack of working L-gulonolactone oxidase (GULO) enzyme would not have mattered; enzymes prolyl

---

[3] Thagard (1999) has a brief discussion of the case of scurvy; although Thagard's 'official' account of mechanism in that work is that a mechanism is 'a system of parts that operate or interact like those of a machine' (p. 106), in discussing scurvy as well as other cases from medicine, he seems to work with a notion of mechanism similar to CM in that he takes mechanisms to be kinds of pathways.

[4] For a useful survey, see Magiorkinis et al. (2011).

hydroxylase and lysyl hydroxylase would have been produced and scurvy would have been prevented.[5]

The history of scurvy is quite interesting. During the Age of Exploration (between 1500 and 1800), it has been estimated that scurvy killed at least two million seamen. Although there were hints that scurvy was due to dietary deficiencies, it was not until 1747 that it was shown that scurvy could be treated by supplementing the diet with citrus fruits. In what is taken as the first controlled clinical trial reported in the history of medicine, James Lind, naval surgeon on HMS *Salisbury*, took 12 patients with scurvy 'on board the Salisbury at sea' (Lind 1753, 149). As he reported, 'Their cases were as similar as I could have them.' The patients were kept together 'in one place, being a proper apartment for the sick' and had 'one diet in common to all'. He then divided them into six groups of two patients each, and each group was allocated to six different daily treatments for a period of 14 days. One group was administered two oranges and one lemon per day for six days only, 'having consumed the quantity that could be spared' (p. 150). The other groups were administered cyder, elixir vitriol, vinegar, seawater, and a concoction of various herbs, all of which were supposed to be anti-scurvy remedies. As Lind put it: 'The consequence was, that the most sudden and visible good effects were perceived from the use of the oranges and lemons; one of those who had taken them being at the end of six days fit for duty.... The other was the best recovered of any in his condition' (p. 150). Lind's experiments provided evidence that citrus fruits could cure scurvy. He said that oranges and lemons are 'the most effectual and experienced remedies to remove and prevent this fatal calamity' (p. 157).

Though Lind had identified a difference-maker, he was side-tracked by looking for the cause of scurvy, which he found in the moisture in the air, though he did admit that diet may be a secondary cause of scurvy (see Bartholomew 2002; Carpenter 2012). But in 1793 his follower, Sir Gilbert Blane, who was the personal physician to the admiral of the British fleet, persuaded the captain of HMS *Suffolk* to administer a mixture of two-thirds of an ounce of lemon juice with two ounces of sugar to each sailor on board. As Blane reported, the warship 'was twenty-three weeks and one day on the passage, without having any communication with the land . . . without losing a man' (quoted by Brown 2003, 222). To be sure, scurvy did appear, but it was quickly relieved by an increase in the lemon juice

---

[5] For a description of the causal pathways of the synthesis of vitamin C in the mammals that can synthesise it, see Linster and Van Schaftingen (2007).

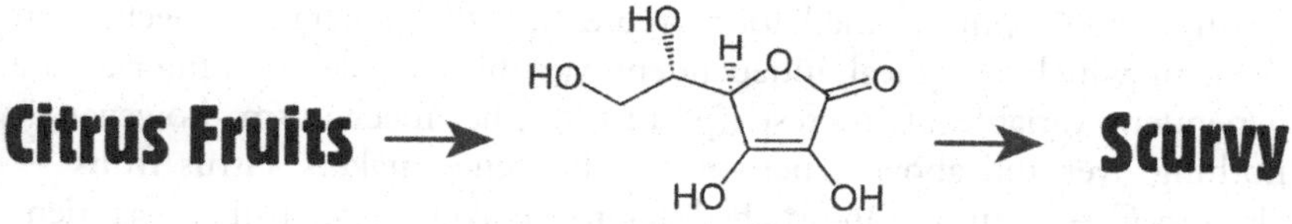

Figure 4.2   The mechanism of scurvy.

ration. When in 1795 Blane was appointed a commissioner to the Sick and Hurt Board, he persuaded the Admiralty to issue lemon juice as a daily ration aboard all Royal Navy ships. He wrote: 'The power [lemon juice] possesses over this disease is peculiar and exclusive, when compared to all the other alleged remedies' (Brown 2003, 222). But even when it was more generally accepted that citrus fruits prevent scurvy, it was the acid that was believed to cure scurvy.

The first breakthrough took place in 1907 when two Norwegian physicians, Axel Holst and Theodor Frølich, looked for an animal model of beriberi disease. They fed guinea pigs with a diet of grains and flour and found out, to their surprise, that they developed scurvy. They found a way to cure scurvy by feeding the guinea pigs with a diet of fresh foods. This was a serendipitous event. Most animals are able to synthesise vitamin C, but not guinea pigs. In 1912, in a study of the aetiology of deficiency diseases, Casimir Funk suggested that deficiency diseases (such as beriberi and scurvy) 'can be prevented and cured by the addition of certain preventive substances'. He added that 'the deficient substances, which are of the nature of organic bases, we will call "vitamins"; and we will speak of a beri-beri or scurvy vitamine, which means, a substance preventing the special disease' (Funk 1912, 342). By the 1920s, the 'anti-scurvy vitamine' was known as 'C factor' or 'anti-scorbutic substance' (see Hughes 1983). In 1927, Hungarian biochemist Szent-Györgyi isolated a sugar-like molecule from adrenals and citrus fruits, which he called 'hexuronic acid'. Later on, Szent-Györgyi showed that the hexuronic acid was the sought-after anti-scorbutic agent. The substance was renamed 'ascorbic acid'. In parallel with Szent-Györgyi's work, Charles King and W. A. Waugh identified, in 1932, vitamin C. The suggestion that hexuronic acid is identical with vitamin C was made in 1932, in papers by King and Waugh and by J. Tillmans and P. Hirsch (see Hughes 1983).

The breakthrough in scurvy prevention occurred when scientists started to look for what has been called the 'mediator' (see Figure 4.2), which is a code word for the 'mechanism', which 'transmits the effect of the treatment to the outcome' (Pearl & Mackenzie 2018, 270). As Baron and

Kenny (1986) put it, mediation 'represents the generative mechanism through which the focal independent variable is able to influence the dependent variable of interest' (p. 1173). This mechanism, however, is nothing over and above a network of difference-makers: citrus fruits → vitamin C → scurvy. One such difference-maker, citrus fruits, was identified by Lind and later on by Blane. This explains the success in preventing scurvy after citrus fruits were administered as part of the diet of sailors. It is noteworthy, however, that Lind and the early physicians did not look for the mediating factor in the case of scurvy. As Bartholomew (2002, 696) notes, Lind did not try to isolate a single common constituent in citrus fruits in particular and in fruit in general which makes a difference to the incidence of scurvy. Instead, he was trying to find out the contribution of difference sorts of vegetable to the relief from scurvy. Still, even without knowing the mediating variable (vitamin C), the intake of citrus in a diet did make a difference to scurvy relief.

In order to find the difference-maker in the case of vitamin C deficiency, it was necessary to find a model (animal) that does not synthesise its own vitamin C. In the late 1920s, Szent-Györgyi and his collaborator J. L. Svirbely used hexuronic acid, recently isolated by Szent-Györgyi, to treat guinea pigs in controlled experiments. They divided the animals into two groups. In one the animals were fed with food enriched with hexuronic acid, while in the other the animals received boiled food. The first group flourished, while the other developed scurvy. Svirbely and Szent-Györgyi decided that hexuronic acid was the cause of scurvy relief and they renamed it ascorbic acid. Ascorbic acid was the sought-after mediating variable: the difference-maker (see Schultz 2002).

We take the case of scurvy to illustrate very clearly that what matters in biological practice when a new pathway is identified are the difference-making relations among the components of the pathway. New mechanists who opt for some theory of causal production in order to ground the causal relations among the components of the mechanism need not, of course, deny the presence of difference-making, nor that difference-making is central in mechanism discovery. The main disagreement between us and new mechanists who adopt causal production is whether difference-making is enough for causation or whether we need to say something more about what *grounds* difference-making. The case of scurvy, while it illustrates our views about difference-making and Causal Mechanism, is not enough to establish the primacy of difference-making. In view of this, why not adopt a productive view of causation? We will come back to this crucial question in Section 4.5.1. In the next section, and to further clarify

CM, we will discuss the consequences of Causal Mechanism for the so-called Russo-Williamson thesis.

4.3.2 Causal Mechanism and the Russo-Williamson Thesis

According to Federica Russo and Jon Williamson (2007), in the health sciences, for a causal connection between A and B to be established, one needs evidence for the existence of both a difference-making relation between A and B *and* a mechanism linking A to B. The thought behind what has been called the Russo-Williamson thesis (RWT) seems straightforward: to accept a causal claim, evidence that a putative cause makes a difference to an effect is not enough, if we do not also know *how* the putative cause brings about the effect, that is, the mechanism (recall Boyle's views on mechanical explanation in Chapter 2).

To illustrate this, consider the following example used by Russo and Williamson to support their thesis. In 1833 it was observed at the Vienna Maternity Hospital that the death rates due to puerperal fever after childbirth in two different clinics of the hospital were different. Ignaz Semmelweis, who was an assistant physician at the hospital, hypothesised that the cause of the difference in the death rates was contamination by 'cadaverical' particles; doctors and medical students examined patients after having carried out autopsies and so they infected them. Semmelweis suggested that handwashing with a solution of 'chlorinated lime' would eliminate the contamination and decrease the death rates. Indeed, after handwashing was introduced, death rates decreased considerably. However, the hypothesis that contamination by 'cadaverical' particles was the cause of puerperal fever was accepted 'only after the germ theory of disease was developed and the Vibrio cholerae had been isolated by Robert Koch, establishing a viable mechanism' (Russo & Williamson 2007, 163). This case, according to Russo and Williamson, shows that 'causal claims made solely on the basis of statistics have been rejected until backed by mechanistic or theoretical knowledge' (p. 163).

Williamson (2011) relies on RWT to raise a problem for both mechanistic and difference-making theories of causation. The problem is supposed to be that these theories, taken on their own, are not compatible with the causal epistemology adopted in biomedicine and other scientific fields, which conforms to RWT. This means that, in biomedicine, as Williamson puts it, 'a causal relation typically signifies the existence of both a mechanistic and a difference-making relation, and evidence of the existence of both the mechanistic relation and the difference-making relation is typically required to establish the causal claim' (p. 435). This argument seems to raise a problem for difference-making accounts of mechanism (and thus for CM). If A causes B in virtue of a mechanism linking A to B, where a mechanism involves a chain of events linked by difference-making relations, it seems that evidence of difference-making is enough to establish a causal claim, contrary to what RWT asserts. In other words, for CM, 'mechanistic' evidence need not be different in kind from difference-making evidence.

So, in claiming that there is a difference between difference-making and mechanistic evidence, Russo and Williamson seem to implicitly reject a difference-making account of mechanism. To be sure, in their 2007 paper, Russo and Williamson do not commit to a particular conception of what a mechanism is. They argue, however, that in the biomedical sciences 'two different types of evidence – probabilistic and mechanistic – are at stake when deciding whether or not to accept a causal claim' (p. 163). In his 2011 paper, Williamson takes cases of absences and double prevention to constitute counterexamples to mechanistic causation, since such cases show that 'a causal relation need not be accompanied by a mechanistic relation' (p. 435). It seems, then, that Williamson had a specific view in mind about what a mechanism is, that is, that mechanistic causation should be viewed in terms of some kind of causal production (in a difference-making account of mechanism, cases of double prevention, at least, would not be problematic; see Woodward 2002).

A difference-making account of mechanism would suggest an alternative picture: while a difference-making relation between A and B is not enough to establish a causal relation between a putative cause A and effect B, a series of difference-making relations between factors causally intermediate between A and B offer further support for the causal claim. But, according to CM, such a causal pathway is exactly what a mechanism is. So, it seems that there is no difference between difference-making and mechanistic evidence for CM, at least of the kind that would lead us to talk about two different *kinds* of evidence.

Williamson and Michael Wilde (2016) deny this view. They assume that there is a distinction between two kinds of evidence, claiming that

> in order to establish that *A* is a cause of *B* there would normally have to be evidence both that (i) there is an appropriate sort of difference-making relationship (or *chain of difference-making relationships*) between *A* and *B* – for example, that *A* and *B* are probabilistically dependent, conditional on *B*'s other causes – , and that (ii) there is an appropriate mechanistic connection (or chain of mechanisms) between *A* and *B* – so that instances of *B* can be explained by a mechanism which involves *A*. (p. 38; emphasis added)

In contrast to this, the case of scurvy shows that looking for mechanistic evidence is just looking for a special kind of 'difference-making' evidence and not for a different kind of evidence. This special difference-making evidence involves looking for the 'mediator'. As we have seen, Lind's experiments provided evidence for a difference-making relationship between citrus fruits and scurvy, but no evidence about how exactly citrus fruits acted so as to prevent scurvy. When it was realised by Funk that scurvy is a 'deficiency disease' – that is, it was produced because of the lack of a particular substance – it became obvious that citrus fruits acted to prevent scurvy by providing that preventive substance. So, scientists started looking for this preventive substance that was the mediating factor between citrus fruits and scurvy. As we have already seen, however, what was required for finding the mediator and establishing the pathway citrus fruits → vitamin C → scurvy was the isolation of a substance (hexuronic acid) from citrus fruits that was such as to prevent scurvy in controlled experiments with guinea pigs by Svirbely and Szent-Györgyi. So, the evidence for identifying the mediator was not evidence about particular entities engaging in activities, or some sui generis type of mechanistic evidence, as one would have believed if the activities-based account of mechanism were true; it was evidence about more difference-making relations, this time between the two initial variables (citrus fruits and scurvy) and the mediating variable vitamin C.

In view of the case of scurvy, we think that RWT can be accepted, but without being committed to the existence of a special type of 'mechanistic' evidence over and above difference-making relations.[6] Moreover, acceptance of RWT does not automatically lead to rejecting a difference-making account of causation. Given a difference-making account of mechanisms, RWT can be understood as follows: typically, to establish a causal connection between A and B, we have to have both evidence for a difference-making relation between A and B and evidence for one or more mediators; but all this evidence is, ultimately, evidence for difference-making relations.[7]

---

[6] Gillies (2011) offers a similar formulation for RWT. He suggests: 'In order to establish that A causes B, observational statistical evidence does not suffice. Such evidence needs to be supplemented by interventional evidence, which can take the form of showing that there is a plausible mechanism linking A to B' (p. 116).

[7] Hill's influential article (1965) has been viewed as offering a version of RWT (cf. Russo & Williamson 2007; Clarke et al. 2014). Note, however, that Hill does not talk explicitly about mechanisms in his paper. He offers 'plausibility' as a criterion for establishing causal claims, which can be understood as the existence of a biologically plausible mechanism; but he does not regard it as particularly important, since '[w]hat is biologically plausible depends upon the biological knowledge

4.4 Inflationary Accounts of Mechanism

Since the emergence of New Mechanism, there had been three dominant accounts of mechanism, offered by Machamer, Darden and Craver (2000); Glennan (1996; 2002) and Bechtel and Abrahamsen (2005). We have presented these accounts in Chapter 2. Other similar general characterisations of a mechanism have appeared in the recent literature, but the following (which Glennan 2017 has called *Minimal Mechanism*) represents a broad consensus:

> A mechanism for a phenomenon consists of entities (or parts) whose activities and interactions are organised so as to be responsible for the phenomenon. (p. 17)

What is important to stress for our purposes is that despite the fact that all these accounts are quite far from traditional mechanistic accounts of causation, they still offer what we might call a 'metaphysically inflated account of mechanism'. In spite of their differences in detail, they are all committed to a certain metaphysics of mechanisms and, in particular, to a certain 'new mechanical ontology', as Glennan has put it (2017, 48). This 'new ontology' of entities, activities, interactions, organisation of parts into wholes and the like creates the further philosophical need – which mechanists try to meet – to explain what they are and how they relate, if at all, with more traditional metaphysical categories.

Recall from Chapter 2 that the following are the key claims of the new mechanists:

1. The world consists of mechanisms.
2. A mechanism consists of objects of diverse kinds and sizes structured in such a way that, in virtue of their properties and capacities, they engage in a variety of different kinds of activities and interactions such that a certain behaviour B or a certain phenomenon P is brought about.
3. The main way to explain a certain behaviour B or a certain phenomenon P in science is to offer the mechanism of it.

---

of the day' (p. 298). As 'strongest support' for causation Hill takes experimental evidence, for example, whether some preventive action does in fact prevent the appearance of a disease. Last, his 'Coherence' criterion involves, among others, establishing a mediator; his example is 'histopathological evidence from the bronchial epithelium of smokers and the isolation from cigarette smoke of factors carcinogenic for the skin of laboratory animals' (p. 298), which was important in establishing a causal connection between smoking and lung cancer.

So, though the new mechanists do not commit themselves to a certain global metaphysics of mechanisms – they do not, for instance, align with the seventeenth-century view of mechanism as configurations of matter in motion subject to laws – their project is not much less metaphysical when they try to offer global accounts of what a mechanism is: the entities might be diverse but they are organised into a mechanism in virtue of their powers, capacities or activities and/or in virtue of their being subjected to laws or at least to invariant generalisations. Mechanisms are typically taken to be things in the world, with more or less objective boundaries, with causally interacting parts bringing about a certain phenomenon P or manifesting a certain behaviour B. Moreover, the blueprint of a mechanistic explanation is decomposition: the behaviour of a system is explained by the interactions/activities of its parts.

Illari and Williamson (2012) have offered their account of mechanism in an attempt to offer a less metaphysically committed view of what a mechanism is, and we have already noted that Glennan's *Minimal Mechanism* is a very similar account. Glennan takes it that an advantage of his minimal account is that mechanisms are everywhere constituting 'the causal structure of the world' (2017, chapter 2). Though we think that both Illari & Williamson and Glennan move the issue in the right direction, we take it that, as it stands, such a minimal account still invites a number of metaphysical questions that, we think, are not relevant to scientific practice. For instance, they invite various metaphysical questions as to the status of entities, their difference from activities, the need to introduce both activities and interactions, the role of the organisation in the performance of the function or behaviour, and so on. These might be philosophically legitimate questions to ask, but, we want to claim, they need not be asked and answered for an understanding of the role of mechanism as a concept-in-use in science.

Could one perhaps read the recent minimal general characterisations of mechanisms in a way that does not emphasise the ontological aspect? We take such characterisations to have two general aims, a metaphysical and a methodological. The metaphysical aim is to specify what exactly mechanisms are *as things in the world*. The methodological aim is to account for the notion of mechanism as this is present in scientific practice. We take it that these two aims are in potential conflict: what is important methodologically in the practice of science may not have clear metaphysical implications, and vice versa. One could choose to emphasise the methodological aspect more than the metaphysical. It's hard, however, to think of the current resurgence of the mechanist world view, even under the 'minimal' characterisations proposed by Glennan and Illari & Williamson, as not having a substantive metaphysical element in it. The very notion of activity as used in the post-MDC literature on mechanisms is heavily metaphysical (being the 'ontic correlate of verbs', as they say). But even if one chose to treat the standard 'minimal' characterisation in an ontologically thinner way (e.g., as not being committed to activities as an irreducible ontological item), it is still the case that even *this* account does exclude certain metaphysical views while fostering others (even if not entailing a specific one); for example, it excludes, on various grounds, Humean accounts of the underlying metaphysics. Reasons such as these make us claim that current accounts of mechanism have significant metaphysical content.[8]

To be more specific, we take the problem with the above accounts of mechanism to be the following. Such accounts invite a number of questions concerning the basic building blocks of mechanisms. For example, consider the claim (which is part of many accounts) that a mechanism consists of two distinct kinds of building blocks: entities (organised in a stable way into a spatiotemporal pattern) *and* activities. But what grounds the difference between entities and activities? Activities are supposed to ground the causal efficacy of mechanisms (according to MDC, mechanisms require 'the productive nature of activities' [2000, 4]); they are the ontic correlates of (transitive) verbs. At issue here, then, is the ontological structure that underlies and unites the scientific theoretical descriptions of mechanisms. Significantly, there are alternative, and competing, ways to characterise this ontological structure: instead of having both entities and activities, one can characterise activities in terms of entities and their causal powers (as, e.g., does Glennan 2017). The question here is: *What is added to scientific practice by insisting that a description of a mechanism has to be couched in some preferred philosophical categories, for example, entities and activities, powers or whatnot?*

New mechanists might reply here that it is unfair to take their metaphysical discussions as intending to inform our understanding of scientific practice. What new mechanists claim, it may be argued, is that the language of mechanisms, entities and activities is descriptively superior compared with the language of objects, properties and laws (or even difference-making relations) in order to characterise the practice of biology. So, MDC write about the 'descriptive adequacy' of their account of

---

[8] A possible exception is Bechtel and Abrahamsen's characterisation of mechanism, which, unlike those offered by Glennan, Illari & Williamson and MDC, is more focused on the methodological aspect and can be read in a metaphysically deflationary way.

mechanisms. The descriptive superiority of the language of entities and activities is then taken to imply a corresponding ontology of entities and activities. Thus, the main question new mechanists ask is something like 'What does scientific practice tell us about ontology?' and not 'What can the new mechanical ontology tell us about scientific practice?'

Concerning the first question, we think it is, by and large, premature and inadequate to infer ontology from language. As a rule, scientific language is compatible with various different metaphysical views (we will come back to this issue in Section 6.8). But even if we let the question from practice to ontology to pass, it would still be imperative to ask the second question too, namely, the question from ontology to practice. To see this let us consider the case of causation. If treating causation in terms of difference-making is enough to understand practice (as the case of scurvy illustrates), then any metaphysically inflated account of causation will add nothing by way of understanding practice further. This means that scientists do not need to commit themselves to any particular account about the metaphysics of causation. So, if the concept of causation that is sufficient for practice is difference-making, then a practice-based view of causation must focus on this (we will come back to this point about causation in Section 4.5.1).

Analogously, a practice-based approach to understanding what a mechanism is ought to focus on concepts that are central in practice, and certainly the notion of mechanisms as causal pathways is central in practice. If new mechanists think that inflationary accounts of mechanisms are central in practice, they ought to show how exactly these metaphysically loaded concepts function within practice, and in particular that they illuminate practice better than the concept of causal pathway. Otherwise, these concepts are not really grounded in practice; they are philosophical additions that have no real function within science.

We do not think, however, that these additions offer further illumination of the practice. Take the case of apoptosis. What clarity (or extra information) would be added to the cytological characterisation of apoptosis offered by Kerr et al. if we were to add that, for example, in the case of the condensation of cytoplasm and nucleus, condensation is an *activity* (whatever that means) and cytoplasm and nucleus are *entities* (whatever that means)? And are the protuberances that form on the surface of the cell and give rise to apoptotic bodies entities or activities? Or, what is the added value of the claim that apoptotic bodies have the *power* to degrade? Similar questions can be asked for the biochemical description of the apoptotic pathway.

In contrast to such metaphysically inflated accounts of mechanisms, our discussion of apoptosis showed that for the identification of a biological mechanism, a theoretical description of the causal pathway (and the function performed) is enough; there is no reason to characterise it further in terms of a preferred ontology. But then, what one can say about what a mechanism in general is, without giving an ontologically inflationary account, is to characterise it as a causal pathway that is theoretically described. Philosophers who characterise mechanisms in general in terms of a preferred ontological inventory have the burden to justify what this further characterisation adds to this deflationary account (i.e., Causal Mechanism) that can be extracted from biological practice.

4.5 Causal Mechanism as a Deflationary Account

The focus on biological practice, then, motivates a deflationary approach to mechanistic talk in biology and biomedicine: such talk does not have to be interpreted in a manner that leads to inquiry into how best to characterise the mechanical ontology of the world. Rather, commitment to mechanism is essentially a methodological (as opposed to an ontological) stance, and mechanism is primarily a methodological concept. We call this stance that incorporates CM as a general characterisation of mechanism *Methodological Mechanism* (MM). In Chapter 10 we will examine MM in detail, taking into account the main results reached in the book. Here we will discuss the main ideas behind CM.

CM can be viewed as based on two main claims, one negative and one positive. The negative claim is that the concept of mechanism as used in scientific practice need not (and should not) be characterised in abstract ontic terms aiming at ontic unity. The positive claim gives a generic characterisation of mechanism as a concept of practice present in various scientific fields. Let's examine these two claims in more detail.

The negative claim differentiates CM from many prevalent philosophical accounts of mechanism. CM remains agnostic (or non-committal) concerning the precise ontology underlying a causal process: it does not commit itself to the existence of activities, powers and the like as distinct from entities and their properties. This deflationary stance is, we think, a decisive advantage of CM over its rivals. First, as a generic account of mechanism applicable to various scientific fields, it retains the advantages of the recent accounts over older reductive approaches to mechanism (cf. Salmon 1997; Dowe 2000). But, second, unlike many recent accounts, it does not read off from scientific practice any views about the ontology of causation (views that do not seem to be supported by the concept of mechanism as used in practice). As noted already, this ontological agnosticism is agnosticism about a 'deeper' ontological description of the mechanism, that is, 'deeper' than the one offered by the relevant theory. As such it is far from implying that mechanisms as causal pathways are not things in the world. According to CM, in elucidating 'mechanism' as a concept of practice we need not go further and commit ourselves to a certain ontological ground of the difference-making relations. Third, by doing so, it avoids the need to answer several metaphysical questions that seem not important in scientific practice. In sum, it takes mechanism to be primarily a methodological, rather than an ontological, concept.

CM is thus opposed to views that, by the very fact of offering mechanistic explanations, biological practice yields specific commitments to a 'mechanistic' ontology. For example, it has been claimed that only a 'local' metaphysics of activities or powers can capture the sense in which mechanisms are both 'real and local' to the phenomenon they produce (Illari & Williamson 2011). While we take it to be fully consistent with CM to view mechanisms as causal pathways that are both 'real and local', in that the identification of a mechanism involves the localisation of the components of the pathway, we resist the further move to a 'local' (a.k.a. lawless) metaphysics. We have two main objections to such a move: first, what one may mean by a 'local' metaphysics is far from clear or uncontroversial (see Chapter 6). But, second, arguments such as those in Illari and Williamson (2011) take place in the context of a discussion on the ontology of mechanisms, and do not directly aim to characterise mechanism as a concept of practice. At the same time, of course, CM is compatible both with the view that the fundamental ontology of the world is broadly neo-Aristotelian as well as with a more Humean view (we will come back to the claim that the local character of mechanisms points to a lawless and powerful mechanical metaphysics in Chapter 6).

The main idea behind the positive claim of CM is that the search for mechanisms improves our understanding of natural phenomena. When scientists look for mechanisms that produce the phenomena, they seek to describe the causal pathways that lead from the initial event of the pathway to the resulting state. A mechanism, then, is some process that shows how exactly the effect is produced.[9] However, to say this is not to make an ontological claim about the structure of the world, but to stress that one of

---

[9] As we have seen in Chapter 2, the idea that mechanistic explanation enhances understanding by identifying a process that connects cause and effect was central in the context of the emergence of

the aims of science should be the discovery of pathways connecting events, which are regular enough so that they can be relied upon to enhance our understanding of how some effects are brought about, and of how we can intervene in order to prevent unwanted outcomes or to treat diseases. Mechanisms, then, can be viewed as theoretically described stable explanatory structures, whose exact content and scope may well vary with our best conception of the world. In particular, a pathway does not have to be described in some privileged (and maybe reductionist) language; pathways in science are described using the theoretical language of the relevant scientific field. Such a description is enough for the identification of a new mechanism, as we have seen in the case of apoptosis. Hence, viewing mechanism in methodological terms licenses our preferred account of mechanism: a mechanism in the biological sciences is a theoretically described causal pathway producing the phenomena of interest.

In the remainder of this section we will first look more closely at how we should understand the metaphysical agnosticism of CM. A main point that we stress is that metaphysical agnosticism is compatible with realism about causal pathways (Section 4.5.1). We will then turn to our claim that CM can serve as a general characterisation of mechanism in the life sciences and argue that it fulfils four important adequacy conditions: it is (1) practice-based, (2) common across fields, (3) topic-neutral and (4) diversifiable (Section 4.5.2).

4.5.1   Causal Mechanism and Metaphysics

We want to resist the temptation to offer a metaphysically inflated account of the causal pathway, in terms of an explicit specification of the mechanical ontology. A key reason for this is that the causal pathway is described in the *theoretical language* of a specific scientific field, and not in some privileged language of ontological categories. This suggests that the form of the description of the mechanism cannot be decided beforehand and in advance of how the concept of mechanism is used. What counts, each time, as a legitimate description of a causal pathway is something that has to be decided by scientific practice. Instead of imposing various ontological categories as those that constitute a general legitimate description of a mechanism, it should be left to the scientists themselves to decide how best to describe mechanisms using the theoretical language they employ to

mechanistic philosophy in the seventeenth century (see, e.g., Boyle's essay 'About the Excellency and Grounds of the Mechanical Hypothesis' in Boyle 1991).

understand and describe the world. CM has the consequence that a series of questions that new mechanists have been concerned with need not concern us if our aim is to understand scientific practice.[10]

However, what about causation? According to CM, causation as a difference-making relation is prior to the notion of mechanism as causal pathway.[11] Does this mean that, in order to identify a mechanism with a causal pathway, it is required to make a commitment about what the ontological nature of causation is? This does not seem necessary for understanding the concept-in-use. As we saw very clearly in the case of scurvy, scientific practice establishes robust causal connections, which can be used for understanding and manipulation, without necessarily being committed to a single and overarching ontic account of causation. Ultimately, whatever fundamental ontological theory of what causation is one might have (e.g., in terms of causal powers or regularities), the identification of causal relationships is based on theory-described difference-making relations; this is what scientists look for when establishing causal relations and causal pathways. In this sense, the causal pathway by means of which a phenomenon Y is brought about by a cause X, given that X initiates a chain of events that leads to Y, is the very network of theory-described difference-making relations among the various intermediaries of X and Y. It is a further question what the truth-makers of these difference-making relations are, and, according to us, this question need not be answered in order to discover and use causal relations in scientific practice. Hence, the point here is that in order to understand what a causal pathway (and hence a mechanism) as a concept-in-use is, and to identify mechanisms, we do not need a theory about the metaphysics of causation: CM is really, ontologically speaking, a deflationary view.[12]

Could perhaps one insist here that the distinction between difference-making and causal production can be found within scientific practice? The argument could go like the following. In biology, inhibition is a feature of many mechanisms. For example, proteins called repressors can bind to DNA and inhibit gene expression. Inhibition, now, is causation by

---

[10] These questions concern, for example, the components and boundaries of mechanisms (cf. Kaiser 2018), the metaphysics of causation (cf. Matthews & Tabery 2018) and the nature of mechanistic levels (cf. Povich & Craver 2018). We will come back to all these issues in subsequent chapters.

[11] It will be the aim of Chapters 5 and 6 to fully substantiate this claim.

[12] There are various other questions that can be raised concerning CM; for example, do we take pathways to be types or tokens? Here again, we defer to practice. Causal pathways, qua things in the world that produce an effect, are concrete particulars. But what is described theoretically in the language of theory is a type of causal pathway.

disconnection, which is considered a problem for productive accounts of causation. The productivity theorist, then, owes us an account of such cases. Similarly, we find in biology and elsewhere *continuous* causes; for example, the cytological description of apoptosis describes a continuous series of stages. It might be argued, with some plausibility, that such cases require a production view of causation, and thus causation as production is properly grounded in scientific practice. It would be a small step from here to arrive at a practice-based metaphysically inflationary account of mechanism.

Yet it's not hard to see that the difference-making account can accommodate difference-makers that are connected by some continuous process. This does not eo ipso imply that difference-making is the ground for productivity; this might well be true but more argument would be required. Rather, the point is that it is one thing to take causation to be production, and quite another to take causation to be a continuous relation among the components of a pathway. Humean accounts can satisfy continuity without being productive.

Our claim that mechanisms are theoretically described causal pathways should not be taken to imply that mechanisms are not things in the world; certainly, causal pathways are as real as anything. Rather, it is meant to highlight that the best description of the causal pathway (and hence of the mechanism) is given by the relevant theory and not by an abstract metaphysical account. Does it follow that every theoretically described pathway is a mechanism? Not at all. It should be a correctly described causal pathway. Not all theoretical descriptions of causal pathways are on a par. First, a theoretical description might fail to capture the actual causal dependencies in the world – such descriptions cannot pick out mechanisms. Second, some theoretical descriptions of a particular causal pathway might be less detailed compared with others, as shown by cytological and biochemical descriptions of apoptosis. How detailed a theoretical description can be is something that will be determined by scientific practice itself. In the case of apoptosis, for example, the initial description was quite detailed: it involved the careful description of various cytological features as well as situating the pathway within the overall functioning of the organism. A less detailed description such as 'cells die on their own', for example, is shown by practice to be too meagre to be of interest. Last, the actual initial identification of a pathway should be seen as the first step that leads to the further elucidation of the various causal dependencies that are involved in the pathway. The end result, as we saw in the case of the biochemical descriptions of the signalling pathways of apoptosis, can be a highly informative theoretical description that embeds the pathway within the known physiological and biochemical functions of the organism.

4.5.2   *The Generality of Causal Mechanism*

Let us now turn to our claim that CM offers a general characterisation of mechanism as a concept-in-use that is common across fields in biology and elsewhere. There are four important adequacy conditions that any such characterisation has to satisfy. First, it has to identify a concept that is actually at work within scientific practice. So, the first condition is that the characterisation has to be *practice-based*. Second, it has to be shown that the notion is *common* across various scientific fields. Third, to be common across many fields, in biology but also in non-biological fields, it has to be as *topic-neutral* as possible. Fourth, it has to be easily *diversifiable* in order to be adapted to more specific contexts without losing the common features that allow us to talk about a shared and common notion. In sum, a general characterisation of mechanism has to be (1) practice-based, (2) common across fields, (3) topic-neutral and (4) diversifiable.[13]

Our examination of the cases of scurvy and, especially, apoptosis and other biological pathways provides ample justification that CM identifies a practice-based notion. As we saw in Chapter 3, an advantage of CM over other accounts is that it can be viewed as a generalisation of a notion that is undoubtedly central in biological practice, that is, the notion of a biochemical pathway. In what follows, we will then focus on conditions (2)–(4).

4.5.2.1   *CM Is Common across Fields*
A possible objection against viewing CM as identifying a notion common across fields is that not all mechanisms are causal pathways. If this is true, then perhaps occurrences of the term 'mechanism' cannot all be accounted for in terms of CM and one should adopt a pluralist stance about what a mechanism in scientific practice is.

So, one might argue that not all mechanisms can be causal pathways, since (at least some) mechanisms are arrangements of entities capable of *implementing* a causal pathway, whether or not the causal pathway is activated; that is, there may exist 'inactivated mechanisms'. Can we

---

[13] Later (in Chapter 10), we will add the condition that such a characterisation has to be non-trivial. But, as this condition is best discussed in the context of our general framework of *Methodological Mechanism*, we will postpone examining it until Chapter 10.

properly describe something as a mechanism, however, even when it is not acting? If, it might be thought, the mechanism of apoptosis can be said to function properly, without apoptosis being initiated, this would seem to be a problem for the present view. Hence, should mechanisms be thought as always acting, or can there be mechanisms waiting-to-act?

Among new mechanists, there is no consensus on this issue. Illari and Williamson (2012), for example, think that a stopped clock would be such a non-acting mechanism; they suppose that examples of cases of 'mechanisms without activities' are cases of 'mechanisms' that instead of producing a change maintain stability (see Illari & Williamson 2012).[14] For Glennan (2017), in contrast, mechanisms are always acting. So, Glennan distinguishes between a mechanical system and a mechanism, where a mechanical system is 'a system that regularly engages in or is disposed to engage in mechanistic processes' (p. 26). A mechanical system, then, is not strictly speaking the mechanism; 'it is rather a thing in which mechanisms act' (p. 26). When a system S does something ψ, 'the mechanism is the organized activities and interactions of entities within the system that is responsible for that ψ-ing' (p. 26) and '[i]f mechanisms truly are understood to be entities acting, the mechanism persists only for the duration of the action' (p. 55). While we disagree with Glennan's general metaphysics of mechanism, we think that this distinction between mechanisms and systems is essentially correct and concordant with biological practice. In the case of apoptosis, the cell (or some part of the cell) can be taken as the system within which the process of apoptosis occurs. But apoptosis itself, qua a mechanism of regulated cell death, is there only when the relevant causal pathway is acting.

In reply, then, to the present objection, we should note that the causal pathway, as we understand it, and as was made vivid with the case of apoptosis, does involve entities, since it is entities that are causally connected by the processes of the causal pathway that leads to the required effect. But to call an arrangement of entities *mechanism*, in our view, is to make a claim about the causal pathway, which according to the theory does yield a certain effect. This does not mean that there are no mechanisms that maintain a stable state instead of producing change. But homeostatic mechanisms in biology are not like stopped clocks, chimneys

---

[14] The clock example was first discussed by Darden (2006), who argues that a stopped clock is a machine and not a mechanism, since mechanisms are 'inherently active'. A 'crystallised form of a molecule' is another example that according to her is not a mechanism, at least in the MDC sense (p. 281).

or pillars supporting roofs, which are examples that Illari and Williamson use to illustrate their view. For instance, the mechanisms within cells that maintain an equilibrium between apoptotic and anti-apoptotic proteins so that apoptosis is prevented involve various causal pathways. In the equilibrium state, anti-apoptotic proteins bind to pro-apoptotic ones, for example; when apoptosis is initiated, the pathways that maintain equilibrium are blocked, and different pathways are activated. In general, homeostatic regulation involves a dynamic equilibrium and so involves change. Thus, a causal pathway need not result in a specific change; its end result may well be the maintaining of a stable state. Of course, given some entities, a causal pathway involving these entities need not be activated; but for the mechanism to cause anything (either a change or a stable state) the causal pathway should be initiated.

Here is a related worry concerning the commonality criterion: Is it justified to generalise from the examples that we have been examining, the mechanisms of cell death (and other biological pathways) and the case of scurvy, to other mechanisms (perhaps all cases of mechanisms) in biology? While further examples will certainly be useful in order to show that the features of mechanisms identified here are typical in the life sciences (and in the following chapters we will discuss various other examples), we think that such a generalisation is (to say the least) very plausible and that cases such as apoptosis and scurvy can be taken as representative. In Chapter 3, Section 3.5, we saw various reasons why apoptosis can be viewed as paradigmatic. In addition to what we said there, we will here discuss two general reasons why we think CM can be generalised.

There are two general reasons to be optimistic that CM can indeed offer a general characterisation of mechanism that captures a concept of practice that is in use in biology and elsewhere. First, our account depends not on very specific features of the particular cases examined but on features that are commonly found in many cases of mechanisms described in biology. That mechanisms correspond to causal pathways, that these pathways are described using the theoretical language of a particular field, that there can be more or less detailed descriptions of causal pathways and that some pathways serve a function (or functions) are very minimal requirements for something to count as a mechanism; for example, it can be easily seen that they apply to all kinds of biochemical (and other) pathways (e.g., signalling pathways, metabolic pathways). This is a main reason why CM can be seen as topic-neutral.

The second reason is that we take the main philosophical contrast to be between CM and metaphysically inflated accounts of mechanism. But if

CM is sufficient to characterise a mechanism such as apoptosis, it is doubtful whether other case studies will make a metaphysically inflated account necessary: if that were the case, to make philosophical sense of the mechanism of apoptosis we would similarly need such a metaphysically inflated account. The reason is that the case of apoptosis exemplifies all the characteristics commonly taken to require a metaphysically inflated characterisation of mechanism: for example, the introduction of both entities and activities to capture both the various macromolecules and their interactions. Hence, if CM is sufficient to understand the present case, it is very doubtful that other cases will require abandoning CM. While we do not want to claim that the prevalent philosophical accounts of mechanisms are ill-motivated, we do claim that they inflate the concept of mechanism *as this is used in science*.

We take it, then, that CM can be generalised to other uses of 'mechanism' within life sciences, but also more broadly: when scientists talk about 'mechanisms' they do so in the context of searching for a process, that is, a *causal pathway*, the identification of which would explain how a particular phenomenon is brought about. If the causal pathway is identified, there is little further interest in understanding it according to a certain theory of causation, or to characterise it in terms of entities bearing powers or engaging in activities, or being involved in activities *and* interactions and the like. In the practice of science, the description of the causal pathway in the *language of theory* is enough for the identification of the mechanism.

4.5.2.2   CM Is Topic-Neutral and Diversifiable

Let us turn now to the criteria of topic-neutrality and diversifiability. To see how it is possible to diversify CM so as to adapt it to specific contexts, consider the relation between the concepts of mechanism and function. In the general accounts presented in Chapter 2, only Bechtel and Abrahamsen explicitly refer to the behaviour of a mechanism in terms of its function. The usual position among new mechanists is that mechanisms are responsible not only for functions but for phenomena in general, where a phenomenon need not be a function.

We agree with this view. If we insist on an account of mechanism broad enough to capture all uses of the concept in life sciences and elsewhere, and given that there are scientific fields where the concept of function is not present (e.g., particle physics, solid state physics, astrophysics, cosmology), an account such as CM seems preferable.[15] Of course, there are contexts

---

[15] Although we are not going to examine in detail examples of mechanisms from these fields in this book, we share the conviction of new mechanists that mechanism is a concept common across

(e.g., in molecular biology) where a mechanism is automatically a mechanism for a certain function; for example, apoptosis is not just a mechanism for cell death in the sense that it leads to cell death, but it also has a homeostatic function within the organism, as we have seen. But it is not the case that a mechanism of star formation, for example, has star formation as its function. The point here is that if we want to claim that a mechanism of cell death and a mechanism of star formation are in some sense the same kind of thing (i.e., that they are both mechanisms) – that is, if we want to give a general account of a mechanism as a concept-in-use across various scientific fields, including non-biological fields – CM (which does not incorporate a robust sense of function) seems a very promising candidate.

At the same time, CM can be easily adapted to capture particular uses of 'mechanism' in various contexts where a specific notion of function is presupposed. This was why we introduced CM-P, as a schema for such more specific versions of CM, in Chapter 3. CM-P, where X is a function that the causal pathway performs, is a typical meaning of what a mechanism is in biological contexts, where a causal pathway is often taken to have a certain function. In the case of apoptosis, the cytological and biochemical descriptions describe the causal pathway, and its role in regulating cell populations identified by Kerr et al. constitutes its function. But even in biology, not all mechanisms have a function in this sense. For example, when necrosis is referred to as a 'mechanism', what is meant is just a causal pathway.

As a further example that shows the prevalence of CM (i.e., causal pathway without a function) within life sciences, take the way pathologists talk about the causes and 'mechanisms of diseases'. They distinguish between causal agents (e.g., viruses), which constitute the *aetiology* of a disease, and the *pathogenesis* of a disease, which concerns the 'mechanism' that leads from the causal agent to a disease state (see Lakhani et al. 2009). We take this to be a clear application of CM. So, when pathologists want to find out how a certain disease is brought about, they look for a specific mechanism, that is, a causal pathway that involves various causal links between, for example, a virus and changes in properties of the organism that ultimately lead to the disease. It is clear, then, that in pathology, to identify a mechanism is to identify a specific causal pathway that connects

___

many scientific fields, and so want our account to be potentially applicable to as many fields as possible.

an initial 'cause' (the causal agent) with a specific result.[16] Other similar examples are the 'mechanisms' of action of drugs; in these cases too there is no function in the robust sense that we saw in the example of apoptosis. Mechanisms of action of drugs and mechanisms in pathology are just causal pathways.

Pathological and pharmacological mechanisms, then, do not have a function as physiological mechanisms do; that is, the effects of pathological and pharmacological mechanisms have not evolved to aid the survival and reproduction of the organism, as the effects of physiological mechanisms have done; but it would be very strange to insist that the notion of mechanism in pathology is very different from the notion of mechanism when applied in contexts when a robust sense of function is assumed (as in the case of apoptosis). According to our account, in pathology and in other biological fields the common core of the notion of mechanism is captured by CM. To that extent, we take CM to be topic-neutral, in the sense that it can be applied to all biological fields, including pathology. But in non-pathological contexts CM-P, where X is a function the causal pathway performs, may better capture the notion implicit in biological practice. To say this is not to abandon CM as a general and common characterisation of a shared notion; it is, rather, to show how a core concept can be diversified to be adapted to specific contexts.

In sum, all mechanisms, even mechanisms such as apoptosis that serve a specific function, are just causal pathways. This means, of course, that all causal pathways are mechanisms. We take this as the core notion of 'mechanism' present in many scientific fields (not only in biology). A mechanism qua causal pathway need not have a function. But, second, in biology we can find a richer notion of 'mechanism', where a mechanism is a causal pathway that has a function within the organism: both apoptosis and necrosis are causal pathways that lead to cell death and so are mechanisms in the sense of CM; but apoptosis does, and necrosis does not, have a function within the organism (i.e., apoptosis regulates cell populations). This is why apoptosis is considered to be biologically interesting or significant by biologists and is taken to be (in contrast to necrosis) a specifically *biological* mechanism. But although there are contexts where a richer notion of mechanism is presupposed (i.e., CM-P), this does not

---

[16] While in pathology and elsewhere there is a distinction between a 'causal agent' and a 'mechanism', strictly speaking when we have a mechanism the cause should be taken not as a single event or an entity but as a whole sequence of them which lead to the effect. However, we could keep the notion of a 'causal agent' to refer to an event or an entity that initiates a causal pathway, and keep the term 'mechanism' for the causal pathway itself.

correspond to a difference in the world: the worldly reference of all kinds of mechanisms are just causal pathways. These considerations, then, provide strong support for the claim that CM identifies a notion of mechanism that is practice-based, common across fields, topic-neutral and diversifiable.[17]

In a sense, then, we agree with the 'received view' among new mechanists about the relationship between mechanisms and functions: mechanisms in biology and elsewhere need not have functions. But new mechanists also typically insist that there is a sense in which a mechanism is *for* a phenomenon, where a mechanism is not just a causal pathway and where the phenomenon is not a function. As we will see in the next chapter, we don't agree with this as we think such a thin reading of mechanism-for is misleading and not particularly helpful. In the next chapter we will introduce a distinction to classify the various conceptions of mechanisms found in the literature, the distinction between *mechanisms-of* and *mechanisms-for*, and will argue that it is best to restrict the category of 'mechanisms-for' to the 'functional' sense of mechanism.

---

[17] The strategy we have followed in this section, where we take a minimal or core concept of mechanism (CM) and then diversify it (CM-P), shares similarities with Glennan's (2017, chapter 5) discussion about types of mechanisms (see also Glennan & Illari 2017, chapter 7). Glennan argues that there are different types of phenomena – some functions, some not – and these suggest different extensions of Minimal Mechanism.

Mechanisms, Causation and Laws

5.1 Preliminaries

This chapter will focus on the concept of mechanism as an ontological category and will examine the relation between mechanisms, causation, laws and counterfactuals. It will place current conceptions of mechanism within the broader context of mechanistic accounts of causation, notably those of Wesley Salmon and John Mackie.

A central distinction that we develop in the chapter is the one between two different conceptions of 'mechanism': mechanisms-of and mechanisms-for (introduced in Chapter 1). *Mechanisms-of* underlie or constitute a causal process; *mechanisms-for* are systems or processes that function so as to produce a certain behaviour. This distinction is used to revisit the main notions of mechanism found in the literature. Examples of mechanisms-of are the various mechanistic accounts of causation. Examples of mechanisms-for are the more recent accounts of the nature of mechanisms, which are based on the ways this concept features in scientific explanation and more generally in scientific practice. According to some new mechanists, a mechanism should fulfil both of these roles simultaneously.

5.2 Mechanisms and Difference-Making

Causal relations are explanatory. If C causes E, then C explains the occurrence of E. Mechanisms are widely taken to be both what makes a relation causal and what makes causes explanatory. So, typically, if one explains the occurrence of event E by citing its cause C – that is, if one asserts that C brings about E or that E occurs because of C – one is expected to cite the mechanism that links the cause and the effect: it is in virtue of the intervening mechanism that C causes E and hence that C causally explains E. On this account of causation, it is not enough to show that E depends on C, where dependence should be taken to be robust, for example, a difference-making relation. Unless there is a mechanism, there is no causation. Difference-making is taken to be enough for prediction and control but not enough for explanation (see Williamson 2011). But, given all this, what is exactly the relationship between mechanisms and causation? Even if we cannot have a causal relation between C and E unless there is a mechanism linking them, something must be said about how exactly a mechanism brings about E, and about how the causal relations within the mechanism itself are to be understood.

There are two ways to view within-mechanism causal relations: either in terms of difference-making or in terms of production. On the former option, C causes E if and only if C makes a difference to E, where the difference-making is typically seen as counterfactual dependence, namely, if C hadn't happened, then E wouldn't have occurred.[1] On the latter option, C causes E if and only if C produces E. 'Production' is a term of art, of course, with heavily causal connotations. The typical way to account for 'production' is by means of mechanism. So, C produces E if and only if there is a mechanism that links C and E. Now, on the production view, mechanisms do not seem to depend on difference-making relations. As according to CM *difference-making* is crucial in understanding mechanism as a concept-in-use, we have to defend our claim that difference-making is conceptually prior to mechanism.

As we will see, some mechanists tend to refrain from using counterfactuals to account for within-mechanism causal relations. For others, counterfactuals are needed in order to ground the laws that characterise the interactions between the components of a mechanism; but counterfactuals may in turn be grounded in lower-level mechanisms (Glennan 2002). Yet other mechanists try to dispense with both counterfactuals and laws, in favour of activities (Machamer et al. 2000). Hence, understanding the relation between mechanisms and counterfactuals requires also clarifying the relation between mechanisms and laws. Laws will thus be central in the argument of this chapter. The key question then, for our purposes, is: Can

---

[1] As is well known, both views face problems and counterexamples. For instance, the production account cannot accommodate causation by absences. The lack of water caused the plant to die, but there is no mechanism linking the absence of water with death. The difference-making account cannot accommodate cases of overdetermination and pre-emption. For instance, suppose that two causes act independently of each other to produce an effect. There is certainly causation, but no difference-making since the effect would be produced even in the absence of each one of the causes (cf. Williamson 2011).

there be an account of mechanism that does not ineliminably rely on some non-mechanistic account of counterfactual dependence?

To be exact, we want to investigate whether a mechanistic theory of causation ultimately relies on a counterfactual theory (and hence, whether it turns out to be a version of the dependence approach) or whether it constitutes a genuine version of the production approach (either because it altogether dispenses with the need to rely on counterfactuals or because it grounds counterfactuals in mechanisms). To be clear on these issues presupposes, as we will argue below, a more careful analysis of the notion of mechanism. Here is the central line of argument we want to investigate: since a mechanism is composed of *interacting* components, the notion of a mechanism should include a characterisation of these interactions, but if (1) these interactions are understood in terms of difference-making relations and (2) these difference-making relations are not in turn grounded in mechanisms, then there is a fundamental asymmetry between mechanistic causation and causation as difference-making, because to offer an adequate account of the former presupposes an account of the latter.

As we will see, not all philosophers who stress the importance of mechanisms for thinking about science are after an account of causation. For some of them, mechanisms are important in understanding scientific explanation and theorising, but it is not the case that causation *itself* is mechanistic (see, e.g., Craver 2007a, 86). Yet even if these philosophers do not have to provide a full-blown account of the ontology of mechanisms, they have to explain the modal force of mechanisms; hence, the issue of the relation between mechanisms and counterfactuals is crucial. So, we can formulate our central question in a more comprehensive way, which will be relevant even for these latter accounts, as follows: given that a mechanism consists of components that interact in some manner, and thus cause changes to one another, does an account of these *interactions* require a commitment to counterfactuals? As we shall see, ultimately, the issue turns not on the need or not to posit relations of counterfactual dependence but on what the suitable truth-makers for counterfactuals are.

5.3   Early Mechanistic Views

5.3.1   *Mackie on Causation and Mechanisms*

J. L. Mackie's (1974) work on causation is the recent common source of both approaches under discussion. In particular, talk of 'mechanisms' in relation to causation goes back to him.

Mackie explicitly appealed to counterfactuals in his definition of the meaning of singular causal statements. He argued that a causal statement of the form '*c* caused *e*' should be understood as meaning '*c* was necessary in the circumstances for *e*', where *c* and *e* are distinct event-tokens. Necessity-in-the-circumstances, he added, should be understood as follows: if *c* hadn't happened, then *e* wouldn't have happened.

Mackie's counterfactuals are not, strictly speaking, true or false: they do not describe, or fail to describe, 'a fully objective reality' (1974, xi). Instead, they can be reasonable or unreasonable assertions, depending on the inductive evidence that supports them (pp. 229–30). For instance, in assessing the counterfactual 'If this match had been struck, it would have lit', the evidence plays a *double role*. It *first* establishes inductively a generalisation. But *then*, 'it continues to operate separately in making it reasonable to assert the counterfactual conditionals which look like an extension of the law into merely possible worlds' (p. 203). So, for Mackie, it is general propositions (via the evidence there is for them) that carry the weight of counterfactual assertions. If, in the actual world, there is strong evidence for the general proposition 'All *F*s are *G*s', we can reasonably assert that 'if *x* had been an *F* it would have been a *G*' based on the evidence that supports the general proposition. Mackie was no realist about possible worlds. He did not think that they were as real as the actual. Hence, his talk of possible worlds was a mere *façon de parler* (p. 199).

These evidence-based counterfactuals *cannot* ground a fully objective distinction between causal sequences of events and non-causal ones. This created a tension in Mackie's overall project. For although he explicitly aimed to identify an *intrinsic* feature of a causal sequence of events that makes the sequence causal, his dependence on evidence-based counterfactuals jeopardised this attempt: whether a sequence of events will be deemed causal will depend, on his view, on an *extrinsic* feature, namely, on whether there is *evidence* to support the relevant counterfactual conditional. It is for this reason that Mackie went on to try to uncover an *intrinsic* feature of causation, in terms of a *mechanism* that connects the cause and the effect.

As Hume famously noted, the alleged necessary tie between cause and effect is not observable. Mackie thought, not unreasonably, that we might still *hypothesise* that there is such a tie, and then try to form an intelligible theory about what it might consist in. His hypothesis is that the tie consists in a 'causal mechanism', that is, 'some continuous process connecting the antecedent in an observed ... regularity with the consequent' (p. 82).

Where Humeans, generally, refrain from accepting anything other than spatiotemporal contiguity between cause and effect, Mackie thinks that mechanisms might well constitute 'the long-searched-for link between individual cause and effect which a pure regularity theory fails, or refuses, to find' (pp. 228–9).

Mackie's own view was that this mechanism consists in the qualitative or structural continuity, or *persistence*, exhibited by certain processes, which can be deemed causal. There needn't be some general feature (or structure) that persists in every causal process. What persists will depend on the details of the actual 'laws of working' that exist in nature. For instance, it can be 'the total energy' of a system, or the 'number of particles' or 'the mass and energy' of a system (pp. 217–18). But insofar as something persists in a certain process, this feature can be what connects together the several stages of this process and renders it causal.

So, early versions of both current views about causation can be found in Mackie's work. In fact, it turns out that the mechanistic view was more central in Mackie's overall approach, since it promised to offer a more objective account of causation and to avoid the notorious context-sensitivity of counterfactual assertions. Yet Mackie's attempt to spell out mechanisms in terms of persistence was deeply problematic.[2]

After Mackie, the counterfactual and the mechanistic approaches parted their ways. They were separately pursued and developed by other able philosophers. The locus of the standard counterfactual theories of causation is the work of the late David Lewis (1986b). Unlike Mackie, Lewis (1973) put forward a *quasi-objectivist* theory of counterfactuals, based on possible-worlds semantics. We will come back to Lewis's theory in Chapter 7 (see also Psillos 2002, chapter 3). Here we will only make the following point, which is relevant to what follows. Lewis's theory renders causation an *extrinsic* relation between events, since it analyses causation in terms of counterfactual dependence among events and it analyses counterfactuals in terms of relations of similarity among possible worlds. In fact, there is a rather important reason why counterfactual theories *cannot* offer an intrinsic characterisation of causation. If causation amounts to counterfactual dependence among events, then the truth of the claim that c causes e will depend on the absence of causal overdeterminers, since if the effect e is causally overdetermined, it won't be counterfactually dependent on any of its causes. But the presence or absence of overdeterminers is certainly *not* an intrinsic feature of a causal sequence.

---

[2] See Psillos (2002, 108–10) for a discussion of the central problems.

5.3.2 The Salmon-Dowe Theory

The locus of the standard mechanistic theories of causation is the work of the late Wesley Salmon. Mackie's preferred account of a causal mechanism in terms of qualitative or structural continuity, or *persistence*, exhibited by certain processes faced significant problems that led Salmon (1984) to argue for an account of causal mechanism that is based on the notion of structure-transference (see Psillos 2002, chapter 4, for a detailed account of Salmon's views). Unlike Mackie, Salmon (1984) tried to characterise directly when a process is *causal*, thereby finding the mechanism that links cause and effect. So, he took processes rather than events to be the basic entities in a theory of physical causation.

Salmon kept the basic idea that '[c]ausal processes, causal interactions, and causal laws provide the mechanisms by which the world works; to understand *why* certain things happen, we need to see *how* they are produced by these mechanisms' (p. 132). But he claimed that the distinguishing characteristic of a causal process (and hence of a mechanism) is that it is capable of transmitting its own structure or modifications of its own structure. Generalising Hans Reichenbach's (1956) idea that causal processes are those that are capable of transmitting a mark, Salmon noted that any process, be it causal or not, exhibits 'a certain structure'. A causal process is then a process capable of *transmitting* its own structure. But, Salmon added, 'if a process – a causal process – is transmitting its own structure, then it will be capable of transmitting certain modifications in the structure' (p. 144).

As many critics noted, however, the very idea of structure-transference (a.k.a. mark-transmission) cannot differentiate causal processes from non-causal ones, since *any* process whatever can be such that *some* modification of *some* feature of it gets transmitted after a single local interaction. A typical example was the shadow of a car with a dent – this is a 'dented' shadow, and the mark is transmitted with the shadow for as long as the shadow is there. In response to this Salmon strengthened his account of mark-transmission by requiring that in order for a process P to be causal, it is necessary that 'the process P would have continued to manifest the characteristic Q if the specific marking interaction had not occurred' (p. 148). It should be clear though that this kind of modification takes us back to persistence! In effect, the idea is that a process is causal if (1) a mark made on it (a modification of some feature) gets transmitted after the point of interaction, and (2) in the absence of this interaction, the relevant feature would have *persisted*, where the required persistence is counterfactual.

Salmon's original promise was for a theory of causation that does *not* involve counterfactuals. The promise, however, was not to be fulfilled. Central to Salmon's theory was the *ability* of a process to transmit a mark. But the ability is a capacity or a disposition, and it is essential for Salmon that it is so. For he wants to insist that a process is causal, even if it is *not* actually marked (p. 147). So, a process is causal if it *could* be marked. Counterfactuals loom large! The message is clear: Salmon's original mechanistic approach cannot do away with counterfactuals. In fact, Salmon's appeal to counterfactuals has led some philosophers (e.g., Kitcher 1989) to argue that, in the end, Salmon has offered a *variant* of the counterfactual approach to causation. Salmon has always been very sceptical about the objective character of counterfactual assertions. So, as he said, it was 'with great philosophical regret' (1997, 18) that he took counterfactuals on board in his account of causation.

Salmon did modify this view further by adopting Phil Dowe's (2000) conserved quantity theory, according to which 'it is the possession of a conserved quantity, rather than the ability to transmit a mark, that makes a process a causal process' (p. 89). On what has come to be known as the Salmon-Dowe theory, a *causal process* is a world line of an object that possesses a conserved quantity. And a *causal interaction* is an intersection of world lines that involves exchange of a conserved quantity.

Dowe fixes the characteristic that renders a process causal and, consequently, the characteristic that renders something a mechanism. A conserved quantity is 'any quantity that is governed by a conservation law' (2000, 91), for example, mass-energy, linear momentum, charge and the like. Dowe's Conserved Quantity theory aims to free the mechanistic view of causation from counterfactuals; but it is far from clear whether this theory can indeed avoid them (for a discussion of this issue, see Psillos 2002, 125–7). However, apart from this problem, the main practical concern is that this account of mechanism is too narrow. For even if *physical* causation – and hence physical mechanism – was a matter of the possession of a conserved quantity, it's hard to see how this account of mechanism can even start shedding any light on causal processes in domains outside physics (biological, geological, medical, social). These will have to be understood either in a reductive way or in non-mechanistic (Salmon-Dowe) terms.

5.3.3   Harré on Generative Mechanisms

A rather liberating conception of causal mechanism was offered by Harré in the early 1970s. Harré connected the traditional idea of power-based causation with the traditional idea that causation involves a mechanism. What he called 'generative mechanism' can be put thus:

generative mechanism = powers + mechanisms

As he put it: 'The generative view sees materials and individual things as having causal powers which can be evoked in suitable circumstances' (1972, 121). And he added: 'The causal powers of a thing or material are related to what causal mechanisms it contains. These determine how it will react to stimuli' (p. 137). For example, an explosion is caused both by the detonation and the power of the explosive material, which it has in virtue of its chemical nature.

On this view of causation, the ascription of a power to a particular has the following form:

X has the power to A = if X is subject to stimuli or conditions of an appropriate kind, then X will do A, *in virtue of its intrinsic nature*.

But this is not a simple conditional analysis of powers, since as Harré stressed, power-ascriptions involve two *analysantia*:

a *specific conditional* (which says what X will or can do under certain circumstances and in the presence of a certain stimulus) and an *unspecific categorical* claim about the nature of X.

The claim about the nature of X is unspecific, because the exact specification of the nature or constitution of X in virtue of which it has the power to A is left open. (Discovering this is supposed to be a matter of empirical investigation.)

It is a fair complaint that, as stated above, the ascription of powers is explanatorily incomplete unless something specific is (or can be) said about the *nature* of the particular that has the power. Otherwise, power-ascription merely states what needs to be explained, namely, that causes produce their effects. This is where mechanisms come in. Specifying the generative mechanism is cashing the promissory note. As Harré put it: 'Giving a mechanism ... is ... partly to describe the nature and constitution of the things involved which makes clear to us what mechanisms have been brought into operation' (1970, 124). So, the key idea in this mechanistic view of causation is this: causes produce their effects because they have the power to do so, where this power is grounded in the mechanism that connects the cause and the effect, and the mechanism is grounded in the nature of the thing that does the causing.

This is a broad and liberal conception of causal mechanism. Generative mechanisms are taken to be the bearers of causal connections. It is in virtue of them that the causes are supposed to produce the effects. But there is no specific description of a mechanism (let alone one that is couched in terms of physical quantities). A generative mechanism is virtually *any* relatively stable arrangement of entities such that, by engaging in certain interactions, a function is performed or an effect is brought about. As Harré explained, he did not 'intend anything specifically mechanical by the word "mechanisms". Clockwork is a mechanism, Faraday's strained space is a mechanism, electron quantum jumps [are]... a mechanism, and so on' (p. 36).

Though this was not quite perceived and acknowledged when Harré was putting forward this conception, this liberal conception of mechanism pointed to a shift from thinking of mechanisms exclusively as the vehicle of causation (what we call *mechanisms-of*) to thinking of mechanisms as whatever implements a certain behaviour or performs a certain function (what we call *mechanisms-for*).

5.4   Mechanisms-for versus Mechanisms-of

Given the views on mechanisms discussed above, as well as the views of new mechanists examined in Chapters 2 and 4, we can say that two traditions have tried to reclaim the notion of mechanism in the philosophical literature of the twentieth century. In exploring the relation between mechanisms and laws/counterfactuals, it is important to distinguish between these two very different senses of 'mechanism'.

The first sense of mechanism is the one typically found in the context of New Mechanism. Central to New Mechanism is the recognition of the fact that scientists try to identify and understand the mechanisms that are responsible for certain phenomena or underlie certain functions, for example, the mechanisms underlying reproduction, gene-transmission, chemical bonding, face-recognition and so on. Mechanisms are always understood as mechanisms *for certain behaviours* (see Glennan 1996; 2002). This has been known in the literature as 'Glennan's law' and is taken to imply that mechanisms are individuated in terms of what they do. For example, there are mechanisms *for* DNA replication or *for* mitosis. What the mechanism does, what the mechanism is a mechanism *for*, determines the boundaries of the mechanism and the identification of its components and operations. Such mechanisms, which we call *mechanisms-for* (i.e., mechanisms *for* certain behaviours/functions), are the mechanisms that, according to new mechanists, figure in explanations in life sciences and elsewhere, and are what many scientists aim to discover. We find mechanisms-for, among others, in MDC (2000), Bechtel & Abrahamsen (2005), Craver (2007a) and Bechtel (2008). This sense of mechanisms is, arguably, the dominant one in various philosophical studies of mechanisms and their role in the various sciences.

The second sense of mechanism is typically found in the context of mechanistic theories of *causation*. These theories, as we have seen in the previous section, aim to characterise the causal link between two events (to fathom Hume's 'secret connexion') in terms of a 'mechanism'. Such theories stem from a general dissatisfaction with standard philosophical views of causation, which fail to explain or take account of the mechanisms by which certain causes bring about certain effects. For this second sense, what the mechanism *does* is not important; what is important is that it is actually there underlying or constituting a certain kind of process. More precisely, what makes a process *causal* is the presence of a mechanism that mediates between cause and effect (or whose parts or moments are the 'cause' and the 'effect'). We call such mechanisms *mechanisms-of*, since they concern the mechanisms of causation. Mechanisms-of are the mechanisms discussed in, for example, theories that view causation as mark transmission (Salmon 1984), persistence, transference or possession of a conserved quantity (Mackie 1974; Salmon 1997; Dowe 2000).[3]

Mechanisms-for are conceived as systems (Glennan 1996; 2002; Bechtel & Abrahamsen 2005) or process-like (Machamer et al. 2000; Craver 2007a; Glennan 2017), performing a function or more generally a behaviour or phenomenon. This is a main difference with mechanisms-of, which are typically conceived as kinds of processes, rather than systems, and need not perform a function. As we have already noted in Chapter 4, the connection between mechanisms and functions is typically not made explicit in general accounts of mechanisms-for (but see Bechtel & Abrahamsen 2005; Garson 2013). New mechanists such as Glennan and

---

[3] Both phrases 'mechanism for X' and 'mechanism of X' are standard locutions in science. Although scientific usage is not entirely consistent on this, there is a tendency to use the phrase 'mechanism *for* X' in biological contexts when X constitutes a certain function and 'mechanism *of* X' more liberally, and also in contexts outside biology, where it seems to mean just a causal process that produces X, without necessarily X being a function. So, there is some justification from scientific practice for using this terminology, as we too take mechanisms-for to be the mechanisms responsible for certain functions. It has been suggested to us that 'productive mechanisms' may be a more accurate term instead of 'mechanisms-of', but we think the term 'mechanism-of' contrasts nicely with 'mechanism-for'. Moreover, although we take this sense of mechanism from mechanistic theories of causation, we use it in a more broad sense. So, 'productive' mechanisms in the technical sense of production are not the only kind of mechanism-of. Causal Mechanism, according to which causation is prior to mechanism and is to be understood in terms of difference-making, is also for us a kind of mechanism-of.

Illari & Williamson who use locutions like 'mechanism for a behaviour' and 'mechanism for a phenomenon' do this precisely in order to have a notion of mechanism-for that is responsible for an X, without the X being necessarily a function.

In contrast to what is perhaps the received view among new mechanists, we do not agree with a non-functional reading of 'mechanisms-for'. If 'mechanism-for' becomes so broad that any causal process can be thought of as a mechanism-for, then we lose the important difference between the two senses of mechanism identified here. But perhaps some new mechanists would insist that 'mechanism-for' should really be understood as a very thin notion; so, whenever we have a causal process and we choose some end point of the process, then we can say that the process is a mechanism *for* that end-state. For, example, if a light beam hits a mirror and gets reflected so as to pass through some particular location, we can say that this process is a mechanism that is responsible for the presence of light at that location[4]. This is partly a terminological issue. The more substantial issue here is that there is a clear distinction between what we call mechanisms-of and mechanisms-for in the functional sense, which is blurred by a very thin reading of 'mechanism-for'.

In order to be conceptually clear, then, we reject the thin reading of 'mechanism-for' and understand all mechanisms-for as mechanisms for certain functions. We think that the thin reading is not really a helpful use of the phrase 'mechanism-for'. For us light reflected by a mirror, for example, is a case of a mechanism-of (i.e., a causal pathway) that is not also a 'mechanism-for'.

The minimal mechanism of Glennan and Illari (2018b), for example, which is defined as a mechanism *for* a phenomenon, can be understood as either a mechanism-of or a mechanism-for, depending on the kind of mechanism to which the definition of minimal mechanism is applied. So, if we apply the definition to the case of the mechanism of apoptosis, which has a particular regulatory function, then we have a mechanism-for. But if we apply the definition to a case like necrosis, where typically there is no function that is performed but what we have is a process that produces a certain end-state, then we have a mechanism-of. In the latter case, even if we characterise necrosis (or any causal process) as a mechanism for a phenomenon, what we really mean is that the mechanism has this

---

[4] Note also here that for someone like Glennan who wants to defend a mechanistic theory of causation, the notion of 'mechanism-for' has to be interpreted as broadly as possible.

phenomenon as its typical effect. What we have in mind in such a case is a mechanism-of.

One natural question may arise at this point. Can a mechanism be *both* what we called a mechanism-for and what we called a mechanism-of? That is, can it be the case that a mechanism *both* underlies or constitutes a causal process *and* is a mechanism for a specific behaviour? Though Harré adopted this view, this position acquired new strength in the early 1990s when Stuart Glennan developed his own mechanistic theory of causation. For him, mechanisms are both what underlie or constitute causal connections between events and thus provide the missing link between cause and effect (mechanisms-of) and at the same time complex systems or processes that are responsible for certain behaviours (mechanisms-for) and are thus individuated in terms of them.

But, as already noted, mechanisms-of are *not* necessarily mechanisms-for. Conceptually, this is obvious if we think of a mechanism as a causal process with various characteristics (such as those discussed above – e.g., they possess a conserved quantity or some kind of persisting structure). There is no reason to think that this kind of mechanism (e.g., the process by means of which the sum of kinetic and potential energy is conserved in some interaction) is a mechanism *for* any particular behaviour. Or consider the reflection of light by a mirror. Although this is clearly a causal process, and so a mechanism-of, there is no sense in which this causal process is *for* something. And of course, if we think of a mechanism as a complex system such that the interactions of its parts bring about a specific behaviour, there is no ipso facto reason to adopt a mechanistic account of causation. In light of this (and if we for the moment restrict the notion of mechanism-of just to mechanisms that feature in mechanistic theories of causation) we arrive at a tripartite categorisation of 'mechanistic' accounts present in the literature (or at *three* different notions of what a mechanism is): mechanisms can be mechanisms-for, mechanisms-of or both.[5]

Where is Causal Mechanism located in terms of this categorisation? For us, the distinction between mechanisms-of and mechanisms-for brings out the contrast between biological mechanisms such as apoptosis that are mechanisms *for* a certain function, and 'mechanisms' in the sense of causes of phenomena. The core sense of mechanism identified by Causal Mechanism says that mechanisms are causal pathways; that is, they are mechanisms-of. But in biological and other functional contexts, the typical notion of mechanism is not CM, but what we called CM-P. So,

---

[5] See Levy (2013) and Andersen (2014a; 2014b) for alternative categorisations.

specifically *biological* mechanisms are mechanisms-for. To return to our main example: both apoptosis and necrosis are mechanisms that lead to cell death, and, hence, they are both mechanisms *of* cell death, but only apoptosis has a function and is thus a mechanism *for* cell death. So, in the case of CM, the mechanisms for/of distinction serves to illustrate the fact that some causal pathways have a function and a functional role to play while other causal pathways do not. The distinction, then, is important for driving home the point that there are mechanisms everywhere (where there are causal pathways) even if there is no function they perform. For example, we can talk about the mechanism of light refraction or the mechanism of beta-decay. So clearly, for us, every mechanism-for is a mechanism-of, but not conversely.[6]

With this map of the conceptual landscape of the philosophical literature on mechanisms in mind, our task now is to examine each case in turn and investigate the relations between each sense of 'mechanism' and laws/counterfactuals. We will briefly look at the Salmon-Dowe theory as the main version of an account of mechanisms-of that are not at the same time mechanisms-for, and then examine in more detail the other two categories, which are the most relevant for our purposes since they represent the main views of what a mechanism is among new mechanists. In Section 5.6 we will discuss mechanisms-for that are not at the same time mechanisms-of. In Chapter 6 we will examine Glennan's theory, in which mechanisms are both mechanisms-of and mechanisms-for.

As noted already, the best known cases of mechanisms-of are those discussed by Salmon and Dowe. Though these accounts of causation are

---

[6] In some cases where scientists refer to 'mechanisms', it may not be clear whether it is the notion of mechanism-of or the notion of mechanism-for (or both) that they have in mind. For example, such uses as 'mechanism of chemical reaction', 'mechanism of speciation' and 'mechanism of action (of a drug)' may in principle be construed in various ways; to insist on a widening of the concept of mechanism-for based on scientific practice (without further argument) seems too quick. On our account, these uses of mechanism are to be viewed as mechanisms-of, that is, as various kinds of causal pathways. Note also what evolutionary biologist Mark Ridley (2004, 35, note 2) writes about the term 'isolating mechanism' in evolutionary biology: 'What is here called an "isolating barrier" has until recently (following Dobzhansky (1970)) usually been called an "isolating mechanism". Some biologists have criticized the word "mechanism" because it might imply that the character that causes isolation evolved in order to prevent interbreeding – that the isolating mechanism is an adaptation to prevent interbreeding.... The use of the term "isolating barrier" is becoming common now, and I follow this usage. However, the older expression could be defended. In biology, a mechanism of X is not always something that evolved to cause X. Compare, for instance, "population regulation mechanism", "mechanism of mutation", "mechanism of speciation", and "mechanism of extinction". Isolating mechanism could mean only a mechanism that isolates, not a mechanism that evolved in order to isolate.' Ridley's two senses of mechanisms clearly correspond to the distinction between mechanisms-for and mechanisms-of.

presented as being compatible with singular causation, it should be quite clear that they rely on counterfactuals. We noted already that in Salmon's account counterfactuals loom large. In fact, counterfactuals play a *double role* in his theory. On the one hand, they secure that a process is causal by making it the case that the process possesses not just an actual uniformity of structure, but also a counterfactual one. On the other hand, they secure the conditions under which an interaction (the marking of a process) is causal: if the marking would have occurred even in the absence of the supposed interaction between two processes, then the interaction is not causal.

On Dowe's account, the very idea of the possession of a conserved quantity for a process to be causal implies that both laws and counterfactuals are in the vicinity. Conserved quantities are individuated by reference to conservation laws, and it is hard to think of a process being causal without the conserved quantity that makes it causal being governed by a conservation law. Counterfactuals are also necessary for Dowe's account of causation, not just because laws imply counterfactuals but also because an appeal to counterfactuals is necessary for claiming that the process is causal. That is, it seems that without counterfactuals there is no way to ground the difference between objects to which conserved quantities may be applied and objects to which they may not. Consider, for example, a single particle with zero momentum versus a shadow with zero quantity of charge; the particle, but not the shadow, is a causal process precisely because it could enter into interactions, which could make its momentum non-zero (see Psillos 2002, 126).

5.5 Mechanisms-for and Mechanisms-of

Let us now turn to an account such as Glennan's, that is, to an account that takes mechanisms to be both mechanisms-for *and* mechanisms-of. There are two parts in Glennan's definition of mechanisms. First, a mechanism consists of components that interact – in this, it is similar to Salmon's account of a mechanism-of as causal process. However, for (early) Glennan the mechanism itself is a system with a stable arrangement of components.[7] In contrast to the view of mechanisms-of as processes (which can in principle be singular causal chains of events), such

---

[7] See Glennan (1996), though in more recent work (Glennan 2010) he drops the stability requirement for some kinds of mechanisms; see Section 5.6.

mechanisms are 'types of systems that exhibit regular and repeatable behaviour' (Glennan 2010, 259).

How should we understand the interactions among the components of such mechanisms? Should they be understood in terms of counterfactuals or not? To answer this question, let us briefly review various possible options.

The first general case we will consider is interactions with laws. Interactions can be governed by laws, where laws are understood in some robust metaphysical sense. For example, according to Fred Dretske (1977), Michael Tooley (1977) and David Armstrong (1983), laws are necessitating relations between universals. So, if there is a necessitating relation between universals A and B, there will be a law between them, and as a result of this law when A is instantiated, so will be B. Suppose we transfer that to the components of a mechanism: when component X instantiates A at some time t_1, some other component Y will instantiate universal B, perhaps at a later time. Or take the rival view (but metaphysically robust too) that laws are embodied in relations between powers. If this is the preferred account of laws, interactions will be understood in terms of powers. Powers are properties possessed by components of a mechanism and produce specific manifestations under specific stimuli. Whereas for Dretske, Tooley and Armstrong the interactions within the mechanism are grounded in the external relation of nomic necessitation, in the powers view, interactions are grounded in the internal relations between the powers of the components of the mechanism. Alternatively, interactions between the components of the mechanism may be viewed as being governed by metaphysically thin laws, for example, by (Humean) regularities. Here, component A can be said to interact with component B in virtue of the fact that this interaction is an instance of a regularity.

If, for the time being, we bracket laws, can we understand the interactions among the components of the mechanism differently? Perhaps counterfactuals can be of direct help here. So, Glennan (2002), following Woodward (2000; 2002; 2003a), understands interactions in terms of change-relating generalisations that are invariant under interventions. Such generalisations are change-relating in the sense that they relate changes in component A to changes in component B. They involve counterfactual situations, in that they concern what would have happened to component B regarding the value of quantity Y possessed by it, if the value of quantity X possessed by component A had changed. These generalisations are invariant under interventions, in that they are about relations between variables that remain invariant under (actual or counterfactual) interventions. These change-relating generalisations, then, are grounded in counterfactuals (called interventionist counterfactuals by Woodward).

But if we are to understand interactions between components in terms of counterfactuals, the next question is: What grounds these counterfactuals? In particular, in virtue of what are interventionist counterfactuals true? The answers here are well known (see, e.g., Psillos 2007). Counterfactuals can be grounded in laws or not. If they are grounded in laws, following what we said previously, these laws can be either metaphysically robust laws of the sort adopted either by Armstrong or by power-based accounts of lawhood, or thin Humean regularities, instances of which are particular token-interactions between components. If the counterfactuals are not grounded in laws, then it's likely that there are counterfactuals 'all the way down', that is, that there are primitive modal facts in the world (see Lange 2009).

In any of these accounts of law-governed within-mechanism interactions, counterfactuals have a central place: either by directly accounting for interactions (as in Woodward's theory), by being part of an account of the nomological dependences that ground the interactions, or as a primitive modal signature of the world.[8] So, if laws regulate the interactions between the components of the mechanism, we cannot do away with counterfactuals in grounding within-mechanism interactions. In Chapter 6 we will consider another option: counterfactuals may be grounded in mechanisms. But for now, let us turn to mechanisms-for.

5.6 Mechanisms-for

So far we have argued that mechanisms-of (mechanisms considered as underlying or constituting causal processes) require laws, and thus difference-making relations must be included in the notion of a mechanism-of. But what about mechanisms-for that view mechanisms as systems or organised entities responsible for certain behaviours? What is the relation between mechanisms-for and laws/counterfactuals? Recall that

---

[8] In Lange's (2009) theory of laws it is a counterfactual notion of stability that determines which facts are lawful and which are accidental. In other theories of lawhood, counterfactuals come 'for free', so to speak, as they must be part of any metaphysically robust theory of laws (such as that of Dretske, Tooley and Armstrong): any such theory must show why laws support counterfactuals. It is not obvious how exactly counterfactuals are part of a regularity view of laws. But note that this is a problem (if at all) for the regularity theorist, and not for the view that interactions have to be understood in terms of laws/counterfactuals. For an attempt to reconcile regularity theory with counterfactuals, see Psillos (2014).

in a mechanism-for the interactions of its parts bring about a certain behaviour or function. A mechanism-for need not commit us to a mechanistic (e.g., à la Salmon-Dowe) account of the causal interactions between its parts.

Here is an argument why mechanisms-for, at least prima facie, have to incorporate laws and/or counterfactuals: a mechanism-for involves components that interact with one another; but laws and/or counterfactuals are needed to account for these interactions; hence, mechanisms-for need to incorporate laws and/or counterfactuals. However, Jim Bogen (2005) has taken the existence of mechanisms that function *irregularly* as an argument against the view that laws and regular behaviour have to characterise the function of mechanisms. In this section we will deal with this argument from irregular mechanisms (in Chapter 6 we will examine the claim that within-mechanism interactions should be viewed in terms of activities and not in terms of laws or difference-making relations).

The first point that we want to stress is that irregular and unrepeatable mechanisms are not as ubiquitous as some philosophers want us to believe. So, consider a claim made by Bert Leuridan (2010). He thinks that mechanisms as complex systems ontologically depend on stable regularities, since there can be no such mechanisms (1) without macrolevel regularities (i.e., the behaviour produced by the mechanism) and (2) without microlevel ones (i.e., the behaviours of the mechanism's parts). Kaiser and Craver (2013, 132) have replied to this that Leuridan's first claim is 'clearly false' since '[o]ne-off mechanisms are mechanisms *without* a macrolevel regularity', where 'one-off mechanisms' are the causal processes discussed by Salmon and others (*mechanisms-of* in our terminology). Moreover, they point to examples where scientists seem to be interested in exactly this kind of mechanism, that is, when they try to explain how a *particular* event occurred (e.g., a particular speciation event).

This kind of reply conflates the two different senses of mechanism we have tried to disentangle. It is not the case that 'singular, unrepeated causal chains . . . are a special, limiting case of [complex system] mechanisms, not something altogether different', as Craver and Kaiser (2013, 131–2) insist. For it is not at all clear that such mechanisms-of are at the same time mechanisms-for, that is, mechanisms *for* a certain behaviour. Similarly, we remain unpersuaded by Glennan's (2010) claim that the mechanisms that produced various historical outcomes are mechanisms-for (he calls them 'ephemeral mechanisms'). In any case, it would be very implausible to insist that any arbitrary causal chain is for a certain behaviour (which is identified with the outcome of the causal chain or, alternatively, with the (higher-level) event constituted by the causal chain). For instance, the reflection of a light ray on a surface is a clear case of a mechanism-of (since it constitutes a causal process), but it is not at all clear that it is a mechanism for a certain behaviour. As we've argued, a thin notion of mechanism-for that can be applied to any causal pathway seems to us misleading and unhelpful.

So, it is not enough to point to singular causal chains in order to argue that there can be irregular mechanisms or one-off mechanisms (mechanisms that function only once) (see also DesAutels 2011; Andersen 2012). Still, one may wonder: Can there be mechanisms-for *without* a corresponding macro-level regularity? Thus, the issue that must be clarified is: What are the conditions for being a mechanism *for* a behaviour? Is it merely to have a function (which is the mechanism's behaviour), or should we, in addition, require that the behaviour be regular?

To make the argument stronger, let us here take function in a wide sense, that is, as not requiring that for something to have a function it has to be the product of conscious design or the result of natural selection (as we have in general assumed in this book). In other words, we are going to take a function in the sense of Robert Cummins (1975), for whom functions are certain kinds of dispositions (see Craver 2001 for such an approach to the functions of mechanisms). In particular, what it is for a mechanism M to have a function F is to have a disposition to F, which contributes to a disposition of a larger system that contains M. Such functions need not be restricted to living systems or artefacts.[9] Yet not anything whatsoever can be ascribed a Cummins-function. In particular, unrepeated causal chains of events, which might well be Salmon's and Dowe's mechanisms-of, need not have a function. We can follow Cummins and say that talk of functions only makes sense when we can apply what Cummins calls the analytical strategy, that is, to explain the disposition of a containing system in terms of the contributions made by the simpler dispositions of its parts.

There can be systems with Cummins functions that exhibit the corresponding behaviour only once; there are many biological functions, the realisation of which requires that the biological entity that has the function cease to exist. An example is the mechanism for apoptosis. Here, the relevant mechanism has a Cummins function; that is, it causally

---

[9] However, note that for some (e.g., Kitcher 1993) we cannot ascribe even Cummins functions to entities that are not products of (either conscious or non-conscious) design, that is, that are not either artifacts or living systems.

contributes to the death of the cell. However, this is a function that, when successfully carried out, can occur only once. But even in such cases, the behaviour of a particular mechanism of apoptosis is a token of *a type of behaviour* that occurs all the time in an organism.

Can there be genuinely *irregular* mechanisms-for, that is, mechanisms-for without a corresponding macrolevel regularity? Bogen (2005) has offered the case of the mechanism of neurotransmitter release as an example of a mechanism-for that behaves irregularly. As this mechanism more often than not fails to carry out its function, there exists no corresponding macro-level regularity; but moreover, and more importantly, Bogen thinks that within-mechanism interactions must themselves be irregular, and thus we must abandon the regularity account of causation in favour of activities.

We do not think that this last conclusion follows from Bogen's example. To see why this is the case, it is useful to distinguish between three cases of what we may call 'irregular' mechanisms-for. The irregularity of mechanisms-for may be only contingent (irregular$_1$), stochastic (irregular$_2$) or (let us assume) more radical (irregular$_3$).

Irregular$_1$ mechanisms-for are mechanisms that could function regularly, but they in fact do not. A defective machine that only functions once in a while is a case in point. Such a machine (1) is a mechanism for a behaviour and (2) functions irregularly. However, it is certainly not the case that a successful operation of the machine is not subject to laws (e.g., laws of electromagnetism, gravity or friction). Nor is it the case that defective machines falsify the regularity account of causation.

Irregular$_2$ mechanisms are like irregular$_1$ mechanisms in that they operate in accordance with laws, but in this case the laws are probabilistic. So, the existence of such mechanisms does not show that within-mechanism interactions are not law-governed, or even that the regularity account of causation is false – regularities can be stochastic. What if the operation of a mechanism is completely chancy, for example, because it involves the radioactive decay of a single atom? Even if we do not have a law here (perhaps because the relevant law concerns a population of atoms rather than a single one), it is not at all clear to us that such a chancy 'mechanism' could be an example of a mechanism-for.[10]

Finally, we can imagine an irregular$_3$ mechanism-for; such a sui generis mechanism operates only once, and its unrepeatability is supposed not to be a contingent matter, but due to the fact that the interactions among its

---

[10] For a notion of 'stochastic' mechanism, see DesAutels (2011) and Andersen (2012).

components cannot *in principle* be repeated. We are not sure that the notion of an irregular$_3$ mechanism-for actually makes sense. But this is the only kind of example we can imagine, where the irregularity or unrepeatability of a mechanism would be a reason to think that its operation is not law-governed. If that's where the friends of genuinely irregularly operating mechanisms can pin their hopes for a non-law-governed account of mechanism, then so be it!

CHAPTER 6

Against Activities

6.1 Preliminaries

This chapter examines a main feature that for several philosophers is, together with entities, an indispensable component of mechanism as an ontological category, that is, activities. According to several new mechanists, activities constitute a novel ontological category, which is required for an account of the productivity of mechanism, and hence has a central place in a mechanistic world view. This chapter offers a critique of the arguments in favour of activities offered in the recent literature. It thus provides a critique against a commonly accepted metaphysics of mechanisms. The main aim of the chapter is to criticise the popular idea that productivity of mechanisms requires a commitment to activities qua a sui generis ontic category.

6.2 Mechanisms and Counterfactuals

Glennan (1996) suggested that *mechanisms themselves* ground counterfactuals. For him, although interactions are understood in terms of interventionist counterfactuals, these counterfactuals are in turn grounded in (lower-level) mechanisms. In this section we first present Glennan's mechanistic theory of causation in more detail (Section 6.2.1) and then focus on his solution to the problem of counterfactuals (Section 6.2.2) and argue that this solution gives rise to what we call the *asymmetry* problem (Section 6.2.3).

6.2.1 Glennan's Early Mechanistic Theory of Causation

Glennan (1996; 2002) took mechanisms to be complex systems (or objects) which bring about a certain activity or are responsible for a certain behaviour. A thermostat might be a stock example of a mechanism in

Glennan's sense. A conventional thermostat works like an on-off switch. A bimetallic coil tips a small mercury-filled glass bottle. The bimetallic coil is made from two different metal strips that have been sandwiched together and then rolled into a coil. As the temperature changes, the two metals expand differently and the coil winds or unwinds. As it does, it tips the glass bottle and the mercury rolls from one end of the bottle to the other. When the mercury falls to one end, it allows an electric current to flow between two wires and the furnace turns on. When the mercury falls to the other end of the bottle, the current stops flowing and the furnace turns off.

According to Glennan (2002, S344):

> (M) A mechanism for a behavior is a complex system that produces that behavior by the interaction of a number of parts, where the interactions between parts can be characterized by direct, invariant, change-relating generalizations.

Mechanisms, he adds, 'are not mechanisms *simpliciter*, but mechanisms *for* behaviors'. For the very same complex system may issue in two different behaviours (e.g., the heart is a mechanism that pumps blood and makes noise). What the mechanism *does* determines its boundaries, its division into parts and the relevant modes of interaction among these parts. Broadly understood, a mechanism consists of some parts (its building blocks) and a certain *organisation* of these parts, which determines how the parts interact with each other to produce a certain output. The parts of the mechanism should be stable and robust; that is, their properties must remain stable, in the absence of interventions. The organisation should also be stable; that is, the system as a whole should have stable dispositions, which produce the behaviour of the mechanism. Thanks to the organisation of the parts, a mechanism is more than the sum of its parts: each of the parts contributes to the overall behaviour of the mechanism more than it would have achieved if it acted on its own. Mechanisms can be contained within larger mechanisms.

There are two major attractions of Glennan's mechanistic theory. The first is that it is descriptively more adequate than the mechanistic approach of Salmon and Dowe. Both of them characterise interactions in terms of the exchange of conserved quantities. To be sure, they do aim at a mechanistic theory of *physical* causation. Still, this account is too narrow to describe cases of causation among higher-level entities. Consider, Glennan says, 'a social mechanism whereby information is disseminated through a phone-calling chain' (p. S346). It is surely otiose and uninformative to try to describe this mechanism in terms of exchange of conserved

quantities. Glennan does not deny that the interactions involved in telephone calls supervene on basic physical interactions. But he is surely right
in saying that we would miss something if we tried to *explain* them in
those terms. We would lose the fact that higher-level interactions form
higher-level kinds. So, Glennan's mechanistic view is broad enough to
account for mechanisms at levels higher than physics. The explanatory
autonomy of higher-level mechanisms is, we think, a lesson that we should
take to heart.

The other attraction of Glennan's mechanistic theory relates to his
demand that understanding causal claims requires knowing what their
underlying mechanisms are (1996, 66). In fact, Glennan wants to make
a stronger point, namely, that 'a relation between two events (other than
fundamental physical events) is causal when and only when these events
are connected in the appropriate way by a mechanism' (p. 56). Although
the weaker claim is very plausible, we don't think the stronger claim is
warranted. Let us examine the relationship between mechanisms and
counterfactuals in Glennan's theory to see the difficulties that arise from
this approach to causation.

6.2.2   Mechanisms as Truth-makers of Counterfactuals

Glennan took his mechanistic approach to offer a rather robust solution to
the problem of counterfactuals. He took laws that are mechanically explicable (in the sense that there is a mechanism that underpins them) to show
in 'an unproblematic way' how 'to understand the counterfactuals which
they sustain' (p. 63). The key idea is that the presence of the mechanism
(e.g., the thermostat) explains why a certain counterfactual holds (e.g., if
the temperature had risen, the furnace would have turned off). Similarly,
the breakdown of a mechanism would explain why certain counterfactuals
fail to hold. In his 2002 paper (see (M) above), Glennan characterised the
interaction of the parts of the mechanism in terms of Woodward's invariant, change-relating generalisations, that is, generalisations that remain
invariant under actual and *counterfactual* interventions.

It seems then that there is a tension between Glennan's views in 1996
and in 2002. According to the earlier view, mechanisms explain via
mechanical laws when certain counterfactuals hold. According to the later
view, certain interventionist counterfactuals explain (or ground) the laws
that govern the interaction of the parts of the mechanism. Consider the
thermostat: it is a mechanical law (ultimately, the law that metals expand
when heated) which explains why it is the case that had the temperature

been higher, the switch would have closed. But why is this a *law*? Because, had we intervened to change one magnitude (e.g., the temperature), the law (that metals expand when heated) wouldn't change and the other magnitude (e.g., the length of the metal strips in the bimetallic coil) would have changed. The tension is obvious: mechanical laws support counterfactuals and counterfactuals render mechanical laws *laws*. Though we are not sure we are faced here with a vicious circle, we are also not sure *where* it can be broken so that the described relation between mechanism and interventionist counterfactuals can get going.

A central and stable feature of Glennan's views is a distinction between the fundamental laws of physics and what he calls mechanically explicable laws. He notes, quite plausibly, that the fundamental laws of physics are *not* mechanically explicable and claims that 'all laws are either mechanically explicable or fundamental, *tertium non datur*' (1996, 61). A mechanically explicable law is a law which is underpinned by a mechanism or, as Glennan says, which 'is explained by the behaviour of some mechanism' (p. 62). He takes it that mechanically explicable laws characterise all the special sciences and 'much of physics itself' (p. 50).

Here is then the resulting picture: interactions among components of a mechanism are governed by laws, which are understood in terms of interventionist counterfactuals; these laws are 'mechanically explicable' – that is, there are other mechanisms that ground them – but these (lower-level) mechanisms themselves contain parts, the interactions among which are understood in terms of counterfactuals, and which are in turn grounded in yet other mechanisms, until we finally reach a level where we run out of mechanisms to explain the laws that govern the interactions among components, and thus to ground the relevant counterfactuals. At this fundamental level, interactions among components are *directly grounded in counterfactuals*. But notwithstanding these not mechanically explicable laws, Glennan insists that at all other levels mechanisms can ground interactions. So, even if we need to introduce counterfactuals to account for interactions, mechanisms seem to have priority over counterfactuals, and thus the account is supposed not to be a version of a difference-making theory of causation, but a genuinely mechanical account.

6.2.3 *The Asymmetry Problem*

If fundamental laws are *not* mechanically explicable, and if they too support counterfactuals (as they do, we suppose), it is not necessary for the truth of a counterfactual that there is a mechanical explanation of it. So, the presence of a mechanically explicable law (and hence of a mechanism) is not a necessary condition for the truth of a counterfactual conditional. Glennan agrees on this; still, he thinks it is a *sufficient* condition. Even if he is right, his theory is incomplete: if some counterfactuals are true even though a mechanism is absent, then there is more to the link between laws and counterfactuals than Glennan's theory admits. Suppose Glennan is right in taking mechanisms to underpin non-fundamental laws. He also subscribes to some kind of supervenience thesis: the non-fundamental laws supervene on the fundamental laws (see 1996, 62 and 66; 2002, S346 and S352). So on Glennan's view, non-fundamental laws are underpinned by mechanisms *and* supervene on fundamental laws, which are not underpinned by mechanisms.

Here is the problem, then. What is the relation between the mechanisms that realise the non-fundamental laws and the more fundamental laws on which the non-fundamental laws supervene? Glennan does not explain. To be sure, he (1996, 66) asserts: 'Although the mechanism responsible for connecting two events may supervene upon other lower-level mechanisms, and ultimately on mechanically inexplicable laws of physics, it is not these laws which make the causal claim true; rather it is the structure of the higher level mechanism and the properties of its parts.'

But this is hardly an explanation of what is going on. One plausible thought is that the fundamental laws govern the interactions of the parts of the mechanism, which realises the non-fundamental law. If this is so, then it would be odd to say that the mechanism that explains, say, Ohm's law is ultimately determined (supervenience *is* a kind of determination) by the fundamental laws that govern the interaction of fundamental particles but that these fundamental laws are *not* (part of) the truth-makers of Ohm's law. Once identified, the mechanism might well have explanatory and epistemic autonomy. But if supervenience holds, the mechanism does not have metaphysical autonomy. We will call this the asymmetry problem.

Here is the problem in more detail: given the existence of not mechanically explicable laws, it is not clear how mechanisms can ground counterfactuals *at any level*. That is, given that the mechanisms at the lowest level depend on counterfactuals, the mechanisms at a level exactly above the fundamental must be equally dependent (albeit *derivatively*) on the fundamental counterfactuals, and so on for every higher level. In other words, to ground counterfactuals at any level, we need the whole lower hierarchy of mechanisms *and* counterfactuals, and since we ultimately arrive at a level where there are either only counterfactuals or only laws (or both), it seems that there is a fundamental asymmetry between mechanisms and laws/counterfactuals. The only way to block the asymmetry would be to argue that we do not need the whole hierarchy in order to ground the counterfactuals at higher levels. Even if this were to be granted for purposes of explanation – that is, even if explanation in terms of mechanisms at level *n* does not require *citing* lower-level mechanisms – metaphysically, the whole hierarchy constitutes the grounds for the mechanism.[1]

We think, then, that the presence of a mechanism is *part* of a sufficient condition for the truth of certain counterfactuals; the fully sufficient condition includes some facts about the fundamental laws that, ultimately, govern the behaviour of the mechanism. This, of course, is entirely consistent with the thought that in most practical situations when it comes to asserting the truth of a certain counterfactual, it is enough to cite the mechanism. The rest of the sufficient condition is not thereby rendered metaphysically redundant, but only explanatorily so.

In sum, given the asymmetry problem and our discussion in Chapter 5, we conclude that if laws are admitted in our notion of mechanism, a reliance on counterfactuals is inevitable. But can we perhaps avoid counterfactuals if we account for within-mechanisms interactions in some other way?

6.3 Activities and Singular Causation

In Chapter 5 we reviewed various options to understand interactions of components of mechanisms, where these interactions are viewed as law-governed. The question now is: Can we have interactions without some notion of law in the background either in terms of regularities or in some more metaphysically robust sense? If yes, then this could be a way to have mechanisms-for without the need to put laws and counterfactuals in the picture.

For some mechanists, the interactions of components have to be understood in terms of *activities*. Activities are a new ontological category that, together with entities, are said to be needed for an adequate ontological account of mechanisms (Machamer et al. 2000; Machamer 2004). Activities are meant to embody the causally productive relations between components. Causation in terms of activities is viewed as a type of singular

---

[1] See Glennan (2011) for an attempt to respond to this argument, and Casini (2016) for a detailed criticism; see also Campaner (2006).

causality, where the causal relation is a local matter; that is, it concerns what happens between the two events that are causally connected, and not what happens at other places and at other times in the universe (as is the case for the regularity theorist). Activities, thus, have been taken to obviate the need for laws.

We disagree. In fact, we are about to argue against the popular activities-based conception of mechanism. We shall start with a criticism of a key argument in favour of activities, that is, that one is led to such a notion if one accepts that causation is *singular* and thus not law-governed.

Does singular causation imply that there are no laws? It would be too quick to infer from singular causal claims that laws are not part of causation. By singular causation we may simply mean that there exist genuine singular causal connections, that is, causal connections between particular event-tokens. But this is not enough to prove that there are no laws in the background. For it is consistent with the existence of singular causal sequences that there are laws under which the causal sequences fall. To use a quick example, on Armstrong's account of laws, singular causation is ipso facto nomological causation since the nomic necessitating relation that relates two universals relates the instances of the two universals too (Armstrong 1997). Interestingly, the same is true if we take singular causation to be grounded in the powers possessed by objects; powers are again *wholly* present in the complex event that constitutes the singular causal sequence. And though there is no nomic relation that relates the two powers, the regular instantiation of the two powers implies the presence of a regularity. So, what both these cases show is that even singular causation can be nomological, that is, subsumed under laws.[2]

Thus, singular causation does not, on its own, constitute an argument in favour of viewing interactions among components of mechanisms as not being law-governed or, more generally, as not depending on difference-making relations. So, friends of activities need to (1) give more reasons to justify the introduction of this new ontological category and (2) explain why activities qua producers of change are themselves counterfactual-free. Although it's conceivable that singular causation just amounts to the local activation of powers which in turn ground activities, powers being

---

[2] There is debate among friends of powers whether such a powers-ontology yields an account of laws in terms of powers (Bird 2007) or a lawless ontology (Mumford 2004). But this need not concern us here.

universals, it's upon the friends of powers to show that we can understand this co-instantiation without also assuming that there is a law present.[3]

We thus reject the widespread claim according to which if causation is singular, this means that mechanisms that embody singular causal relations are not grounded on laws. As we showed, causation can be singular *and* nomological at the same time. Let us now see what other reasons new mechanists have given in favour of an activities-based conception of mechanisms.

6.4 Against Activities I: The MDC Account

As we have already seen, Machamer, Darden and Craver claim: 'Mechanisms are entities and activities organised such that they are productive of regular changes from start or set-up to finish or termination conditions' (2000, 3). On the face of it, the MDC characterisation of a mechanism is fairly similar to Glennan's. On closer inspection, there is a central difference. MDC introduce the concept of *activity* as a means to account for the interaction between the parts of the mechanism and its overall causal efficacy. The MDC approach is exciting, especially when it comes to the detailed description and classification of how mechanisms are taken to operate in neurobiology. But for the purposes of this chapter, we will examine only the notion of activity. This notion is central to MDC's mechanistic view of causation since, as they say, 'activities are types of causes' (p. 6) and 'activities are needed to specify the term "cause"' (p. 8).

As we see it, their view is that an adequate understanding of the concept of mechanism requires an *ontological* shift: we need to accept the existence of activities on top of the usual commitments to entities, properties and processes. This unparsimonious move is recommended on the basis of their claim that mechanisms are 'active': 'they do things' (p. 5). They think that unless activities are accepted as ontological bedfellows of entities, properties and processes, mechanisms will be *passive*: things might be done *via* them, but not *because of* them. They also claim that appeals to causal

---

[3] See Waskan (2011) for a mechanist account of the contents of causal claims that is not based on counterfactuals and Woodward's (2011) answer that causation as difference-making is fundamental in understanding mechanisms; Menzies (2012) provides an illuminating account of mechanisms in terms of the interventionist approach to causation within a structural equations framework; last, Glennan (2017, chapters 5 and 6) offers a detailed treatment of mechanistic causation as a productive account of causation not reducible to difference-making relations. We will examine Glennan's account in detail in Section 6.7.

laws, or to invariant generalisations, fail to capture the productivity of a mechanism, which 'requires the productive nature of activities' (p. 4).

MDC's 'dualism', as they put it, requires that there is a fine distinction between entities (with their properties) and activities. But is there? As is usual in philosophy, we are first given some examples. So, cases such as bonding, diffusion, depolarisation, attraction and repulsion are cases of *activity*. But what do all these share in common in virtue of which they are *activities*? What we are told is that 'activities are the producers of change' (p. 3). But production is itself an activity. So, we are not given an illuminating account of that which some things share in common, in virtue of which they are activities.

MDC say the following of the relation between entities and activities: 'Entities and a specific subset of their properties determine the activities in which they are able to engage. Conversely, activities determine what types of entities (and what properties of those entities) are capable for being the basis of such acts.... Entities and activities are correlatives. They are interdependent' (p. 6). It follows that entities and activities are ontically on a par: they determine each other. They say this more explicitly when they claim that '[t]here are no activities without entities, and entities do not do anything without activities' (p. 8).

We think the supposed ontic parity between entities and activities is wrong-headed. First, it's conceivable that there are entities without activities. Indeed, there may be entities capable of engaging in certain activities, but the prevailing circumstances or the laws of nature may be such that they *fail* to engage in these activities. (If what matters is the ability of an entity to engage in an activity and not the actual occurrence of this activity, then it is clear that MDC have to rely on counterfactuals to illuminate the link between entities and activities.) Second, we cannot see how activities can determine what *types* of entities can engage in them. There may well be an open-ended list of types of objects that can engage in some activity, and they may share very little, if anything, in common. Take the activity of *playing*. It's hard to say that it determines what kinds of entities (and what properties) are involved in this activity. Admittedly, this is a case of a highly generic activity and it might be problematic precisely because of this. There are cases of more specific activities, where the activity is performed by certain *types* of objects. It then might *seem* that the activity does determine what types of object can engage in it. An example of such a specific activity might be the activity of *pushing*. It seems that this activity determines that the objects involved in it must have certain properties, for example, rigidity, bulk and so on. But we think appearances are deceptive.

Epistemically, we might first classify a certain type of activity and then identify what kinds of objects engage in it. But from this it does not follow that this is the order of ontic dependence too. On the contrary, objects can engage in certain activities *because* they have certain properties, and not the other way around.

Consider the activity of chemical bonding. Does this activity determine that entities that engage in it must have a certain electronic structure? Not really. Chemical bonding could not exist without some entities having the right electronic structure. So, not only are the latter presupposed ontically for the activity, but they also fully *determine* this activity: the activity of bonding *consists* in the fact that certain entities with certain electronic structure behave in a certain way when they are in proximity. The dependence of the activity on the properties of entities becomes clear when the activity *fails* to take place. Consider the case where chemical bonding does not take place, for example, the case of noble gases. There, you have the entities without the activity of chemical bonding precisely because the entities and their properties determine that a certain activity *cannot* take place. The situation is exactly symmetrical when the activity *does* take place.

The conclusion we draw is that activities cannot be ontically on a par with entities. But one may wonder: Why should MDC want to hypostatise activities? Why isn't it enough to talk in terms of entities and their properties? MDC may be right in protesting against process-theorists that entities are indispensable in understanding mechanisms; they may rightly claim that the programme of reducing entities to processes is 'problematic at best'. But they also want to argue against 'substantivalists', that is, those who 'confine their attention to entities and properties, believing that it is possible to reduce talk of activities to talk of properties and their transitions' (p. 4). Against them, MDC claim that entities and their properties are not enough for the characterisation of mechanisms: activities are also required. Now, the substantivalists that MDC have in mind take the properties of the entities to be dispositional; they equate them with *capacities* or *active powers*. This is a quite powerful ontology. The friends of active powers would surely protest that given that active powers are granted to entities, talk of activities as *distinct* from these powers is redundant.[4]

---

[4] Consider how Harré (2001, 96) understands an active power: 'a native tendency or inherent capacity to act in certain ways in the appropriate circumstances'. Activities come for free if Harré is right. Note that Harré too favours a mechanistic account of causation.

MDC offer two arguments for activities on top of capacities. We think they are both problematic. The first argument is this:[5] '[I]n order to identify a capacity of an entity, one must first identify the activities in which that entity engages' (p. 4). Even if right, this is irrelevant. It only raises an epistemic point: we cannot know what capacities an entity has unless we first know what it *does*. From this, it does not follow that activities are ontically on a par with capacities. Nor does it follow that it is not the capacities of an entity that determine what activities it engages in. Quite the contrary. To use their own example, it is *because* aspirin has the capacity to relieve headaches (a capacity which we take it to be grounded in its chemical composition) that aspirin engages in this activity, that is, headache- relieving. If capacities are granted, then activities supervene on them. And this remains so, even if, from an epistemic point of view, we need to attend to the (observed) activities in order to conjecture about the capacities.

The second argument that MDC offer is this: '[S]tate transitions have to be more completely described in terms of the activities of the entities and how those activities produce changes that constitute the next change' (p. 5). Here the emphasis is on the *production*. As they explain, activities add the 'productivity' by which changes in properties (state-transitions) are effected. But isn't this question-begging? Many would just deny that there is anything like a productive continuity in state transitions. All there is, they would argue, is just regular succession (or some kind of dependence). In any case, the friends of capacities would argue that there is productive continuity in state-transitions, but that this is grounded in the natures of the entities engaged in state transitions. If water has the capacity to dissolve salt, and if this capacity is grounded in the natures of water and salt, then all that is needed for the dissolution of salt in water (i.e., the activity) is that the circumstances are right and the two substances are brought into contact.

We have a final, but central, objection to MDC: they cannot avoid counterfactuals. Counterfactuals may enter at two places. The first is the activities themselves. Activities, such as bonding, repelling, breaking or dissolving, are supposed to embody causal connections. But one may argue that causal connections are distinguished, at least in part, from non-causal ones by means of counterfactuals. If 'x broke y' is meant to capture the claim that 'x *caused* y to brake', then 'x broke y' must issue in a counterfactual of the form 'if x hadn't struck y, then y wouldn't have broken'. So

---

[5] The essence of this argument is repeated in Machamer (2004).

talk about activities is, in a sense, disguised talk about counterfactuals. The second entry point for counterfactuals is the characterisation of interactions within the mechanism. We have already seen (in Section 6.2) Glennan insisting that this interaction should be captured in terms of the invariance of the relationships among the parts of the mechanism under actual and counterfactual interventions. MDC are not quite clear on what the interaction within the mechanism consists in. Note that it wouldn't help to try to explain the interaction between two parts of a mechanism (say, parts A and B) by positing an intermediate part C. For then we would have to explain the interaction between parts A and C by positing another intermediate part D and so on (ad infinitum?).

We take this to be a crucial problem of the mechanistic approach to causation. In a sense, this approach fills in the 'chain' that connects the cause and the effect with intermediate loops. But there is still no account of how the loops interact. Here, it might well be the case that the most general and informative thing that can be said about these interactions is that there are relations of counterfactual dependence among the parts of the mechanism. Even if we posited activities, as MDC do, we would still need counterfactuals to make sense of them, as we have just seen. In any case, if we are right, there is more to causation than mechanisms.

6.5 Against Activities II: Glennan's Approach

Let us now turn to Glennan's approach to activities, which we will characterise as a top-down approach, partly in order to contrast it with Illari and Williamson's bottom-up approach that we will examine next. Note that, by a top-down approach we do not mean an approach that starts with a preferred ontology and tries to understand science on the basis of it. No new mechanist uses such an approach. Rather, all start with science, but in trying to understand scientific practice new mechanists use metaphysical notions in various degrees and differ in their interest to develop a systematic metaphysical picture. So, for example, some new mechanists may be suspicious of metaphysics and more interested to examine how mechanisms function within science.[6] Others, like Glennan, are interested in developing a systematic mechanical metaphysics. The top-down/bottom-up distinction applies to new mechanists who fall in this latter camp and illuminates, it seems to us, a difference in methodology.

---

[6] Bechtel is such an example of a new mechanist (although see Craver & Bechtel 2007).

We take our distinction between top-down and bottom-up approaches not to be extremely sharp, but a matter of degree or emphasis. Still, a difference of degree is a difference: some new mechanists focus more on developing a systematic and coherent metaphysical picture than others. So, this new mechanical strategy is not top-down as was, for instance, in the case of someone like Descartes (and his *Principia*). But note that even in the case of Descartes, where the top-down approach is prominent, there is also a focus on methodological issues; so, the resulting Cartesian image, where metaphysics has primacy, can be seen as a reconstruction of a network of philosophical practices that puts the emphasis on metaphysics and not on the methodology, just as in the case of some new mechanists. We think that this similarity justifies calling such an approach 'top-down', even if it is fully naturalistic.

As we have already seen, Glennan has recently put forward what he calls Minimal Mechanism: 'a mechanism for a phenomenon consists of entities (or parts) whose activities and interactions are organised in such a way that they are responsible for the phenomenon' (Glennan 2017, 13). Though minimal, this account is 'an expansive conception of what a mechanism is' (p. 106), mostly because it involves commitment to activities as a novel ontological category. 'Activities', Glennan claims, 'cannot naturally be reduced to properties of or relations between entities' (p. 50).

Here then are some characteristics of activities, according to Glennan. Activities are concrete: 'they are fully determinate particulars located somewhere in space and time; they are part of the causal structure of the world' (p. 20). Activities are the ontic correlate of verbs. They include anything from walking to pushing to bonding (chemically or romantically) to infecting. Given this, activities 'are a kind of process – essentially involving change through time' (p. 20). Some activities are non-relational (unary activities) since they involve just one entity, for example, a solitary walk. But some activities involve interactions: they are non-unary activities, namely, activities that implicate more than one entity (p. 21).

Most activities, Glennan says, 'just are mechanistic processes', that is, spatiotemporally extended processes that 'bring about changes in the entities involved in them' (p. 29). What, then, is a *mechanistic* process? According to Glennan, 'To call a process mechanistic is to emphasise how the outcome of that process depends upon the timing and organisation of the activities and interactions of the entities that make up the process' (p. 26).

Now, it appears that there is a rather tight circle here. A process is mechanistic when the entities that make it up engage in *activities*. But if activities just are *mechanistic processes*, then a process is mechanistic when the entities that make it up engage in mechanistic processes. Not much illumination is achieved. Perhaps, however, Glennan's point is that activities and processes are so tightly linked that they cannot be understood independently of each other. Yet there seems to be a difference: activities (are meant to) imply action. To describe something as an activity is to imply that something acts or that an action takes place. A process need not involve action. It can be seen as a (temporal or causal) sequence of events. In fact, it might be straightforward to just equate the mechanism with the process, namely, the causal pathway that brings about an effect. In the sciences all kinds of processes are characterised as mechanistic irrespective of whether they are 'active' or not.

Let us illustrate this point by a brief discussion of the case of active versus passive membrane transport, which are the two mechanisms of transporting molecules across the cell membrane. The transportation of the molecules takes place across a semi-permeable phospholipid bilayer and is determined by it. Some molecules (small monosaccharides, lipids, oxygen, carbon dioxide) pass freely the membrane through a concentration gradient, whereas other molecules (ions, large proteins) pass the membrane against the concentration gradient and use cellular energy. The main difference between active and passive transport is precisely that in active transport the molecules are pumped using ATP energy, whereas in passive transport the molecules pass through the gradient by diffusion or osmosis. These different mechanisms play different roles. Active transport is required for the entrance of large, insoluble molecules into the cell, whereas passive transport allows the maintenance of homeostasis between the cytosol and extracellular fluid. But they are both causal processes or pathways, even though only one of them is 'active'.

Glennan (2017, 32) takes it that 'the most important feature of activities' is that most or all activities are mechanism-dependent. This, he thinks, suggests that 'the productive character of activities comes from the productive relations between intermediates in the process, and that the causal powers of interactors derive from the productive relations between the parts of those interactors'. But this is not particularly illuminating. Apart from the fact that production is itself an activity, to explain the productive character of activities by reference to the productive activity of intermediaries or of the constituent parts of the mechanism just pushes the issue of the productivity of an activity A to the productivity of the constituent activities $A_1, \ldots, A_n$ of the mechanism that realises A. Far from explaining how activities are productive, it merely assumes it. Now,

Glennan takes an extra step. He takes it that some producings are explained 'in terms of other producings, not in terms of some non-causal features such as regularity, or counterfactual dependence' (p. 33). In the context in which we are supposed to try to understand what distinguishes activities from non-activities, this kind of argument is simply question-begging.

If what makes entities engage in activities are their properties and relations to other entities, in what sense are activities things distinct from them? In what sense are activities 'a novel ontological category'? Here, we find Glennan's argument perplexing. His chief point is that thinking of activities as fixed by the properties and relations of things 'reduces doing to having; it takes the activity out of activities' (p. 50). The language of relations 'is a static language' (p. 50). But activities, we are told, are 'dynamic' (p. 51).

Let us set aside this figurative distinction between doing and having. After all, it is by virtue of having mass that bodies gravitationally attract each other, according to Newton's theory of gravity. More generally, it is by virtue of having properties that things stand in relations to each other, some of which are 'static', for example, being taller than, while others are 'dynamic', for example, being attracted by. To see why activities do not add something novel to ontology, let us stress that for Glennan activities are fully concrete particulars: 'Any particular activity in the world will be fully concrete, though our representations of that activity may be more or less abstract' (pp. 95–6). Now, if activities are always particular, and if they are always specific, like pushings, pullings, bondings, infectings, dissolvings, diffusings, pumpings and so on, there is no need to think of them as comprising a novel ontic category. For each fully concrete activity, there will be some account in terms of entities, their properties and relations. A pushing is an event (or a process) that consists in an object changing its position (over time) due to the impact by another body. Indeed, the very event itself *consists* in a change of the properties of a thing (or of its relations to other things). Similarly, for other concrete activities: there will always be some description of the event or the process involved by reference to the changes of the properties of a thing (that engages in the 'activity') or of the relations with other things.

Take the case of a mechanism such as the formation of a chemical bond. Chemical bonding refers to the attraction between atoms. It allows the formation of substances with more than one atomic component and is the result of the electromagnetic force between opposing charges. Atoms are involved in the formation of chemical bonds in virtue of their valence electrons. There are mainly two types of chemical bonds: ionic and covalent. Ionic bonds are formed between two oppositely charged ions by the complete transfer of electrons. The covalent bond is formed by the equal sharing of electrons between two bonded atoms. These atoms have equal contribution to the formation of the covalent bond. On the basis of the polarity of a covalent bond, it can be classified as a polar or non-polar covalent bond. Electronegativity is the property of an atom by virtue of which it can attract the shared electrons in a covalent bond. In non-polar covalent bonds, the atoms have similar electronegativity. Differences in electronegativity yield bond polarity. In *describing* this mechanism, there was no need to think of particular activities as anything other than events (sharing of electrons) or processes (transfer of valence electrons) that are fixed by the properties of atoms (their valence electrons; electronegativity) and the relations they stand to each other (similar or different electronegativity).

Glennan, however, takes it that 'processes are collections of entities acting and interacting through time' (p. 57). Elsewhere (p. 83), he notes that a mechanism is a 'sequence of events (which will typically be entities acting and interacting)'. If we were to follow Bishop Berkeley's advice to '*think* with the *learned* and *speak with the vulgar*', we could grant this talk in terms of activities, without hypostatising activities over and above the properties and relations by virtue of which entities 'act and interact'. We conclude that 'activity' is an abstraction without ontological correlate.

When he talks about entities, Glennan takes it that a general characteristic of entities is this: 'The causal powers or capacities of entities are what allow them to engage in activities and thereby produce change' (p. 33). What produces the change? It seems Glennan's dualism requires that there are causal powers *and* activities and that the former enable the entities that possess them to engage in activities, thereby producing changes (to other entities). It's as if the activities exist out there ready to be engaged with by entities having suitable causal powers. Glennan is adamant: 'activities are not properties or relations; they are things that an entity or entities do over some period of time' (p. 96).

But this cannot be right. The activities cannot exist independently of the entities and their properties (whether we conceive them as powers or not), as Glennan himself admits. What activities an entity can 'engage with' depends on the properties of this entity. Water can dissolve salt but not gold, to offer a trivial example. The 'activities' an entity can engage in are none other than those that result from the kind of entity it is. If you assume powers, as Glennan does, then the activities of an entity are fixed by the manifestation of its powers (given suitable circumstances). Given a power ontology, the powers are the producers of change; the activities are merely the manifestation of powers.

As Glennan admits: 'The central difference between activities and powers is that activities are actual doings, while powers express capacities or dispositions not yet manifested' (p. 32). As just noted, assuming particulars with powers, activities are the manifestation/exercising of these powers. When a cube of salt is put in water, it dissolves. The dissolving is the manifestation (assuming a power-ontology) of the active power of water to dissolve (water-soluble) materials and the passive power of the salt to get dissolved. The dissolving takes time (and hence it is a process); but it is not acting in any sense; it does not produce any changes in the salt; it *consists* in the changes in the salt. The 'scraping of the skin off the carrot' (Glennan's example) *is* the removal of the skin of the carrot (at least on this particular occasion) and hence it does not cause (or produce) the removal. Activities do not produce anything; they *are* the productions (of effects).

6.6   Against Activities III: Illari and Williamson's Approach

While Glennan's motivation for activities comes from the metaphysics of mechanisms, other philosophers vouch for activities on the grounds that science requires them. The general motivation appears to be that science must constrain metaphysics. Not only is it the case that what there is has to be compatible with what science describes, but also the best route to the fundamental structure of the world should be the descriptions that science offers. Thus, proponents of activities have argued that if we take seriously the descriptions offered in such fields as molecular biology or neurobiology, we find that activities are central in these descriptions (Machamer et al. 2000; Illari & Williamson 2013). Illari and Williamson, in particular, think that '[t]here is a good argument from the successful practice of the biological sciences for the appeal to activities in the characterisation of a mechanism' (Illari & Williamson 2013, 71).

Illari and Williamson (2011) offer a bottom-up argument in favour of what they call an 'active metaphysics' for the workings of mechanisms, by which they mean a metaphysics in terms of capacities (cf. Cartwright 1989), of powers (cf. Gillett 2006) or of activities (cf. Machamer et al. 2000). They contrast active metaphysics with 'passive' metaphysics, which characterises the working of mechanisms in terms of laws or counterfactuals. In this section we examine this kind of bottom-up argument, which we are going to call the 'local argument'.

Although we are here treating the local argument as an argument in favour of activities, Illari and Williamson take the argument to be more general, as it does not differentiate between activities-based and power-based views. In fact, Illari and Williamson (2013) offer reasons to prefer an ontology based on entities and activities over an ontology based on entities and capacities, a main reason being that an ontology of activities is more parsimonious. But since these arguments are largely metaphysical, and we are here focusing on bottom-up arguments, we are going to examine the local argument in its general form.

Illari and Williamson argue that biological practice and, in particular, the fact that mechanisms are taken to be explanatory, constrains the ontology of mechanisms. More specifically, they think that a metaphysics of mechanisms that views within-mechanism interactions in terms of laws or counterfactuals is 'in tension with the actual practice of mechanistic explanation in the sciences, which examines only local regions of spacetime in constructing mechanistic explanations'. So, passive approaches do not 'allow mechanisms to be real and local . . . [O]nly active approaches give a local characterisation of a mechanism' (Illari & Williamson 2013, 835). They think, then, that the local argument establishes that a characterisation of mechanism has to be given in terms of an active metaphysics and not in terms of 'counterfactual notions grounded in laws or other possible worlds' (p. 838).

The local argument can be reconstructed as follows:

The practice of mechanistic explanation requires that mechanisms be local (1).
This in turn implies that a characterisation of mechanism has to be local (2).
But only a metaphysics of powers or activities is a local metaphysics (3).
So, a local characterisation of mechanism requires a metaphysics based on powers or activities (4). (pp. 834–8)

In response to this argument for an 'active' metaphysics of mechanisms, it seems to us that 'local' cannot have the same meaning in premises (1) and (2), on the one hand, and in premise (3), on the other: we can have local mechanisms without a local metaphysics. There are three points to note here.

First, it is certainly true that mechanisms are local to the phenomena they produce. In this context, 'local' means that mechanistic explanation involves the localisation of the parts into which the mechanism is decomposed, the operations of which produce the phenomenon for which the mechanism is responsible. Indeed, as Bechtel and Robert Richardson (2010) have argued, localisation is a central strategy in constructing a mechanistic explanation: scientists decompose the phenomenon under study into component operations, and 'localise them within the parts of the mechanism' (p. xxx). But then, localisation of parts can fully capture the sense in which mechanisms are 'local', without entailing a 'local' metaphysics, which is supposed to underlie a characterisation of the interactions among components, and not only the components themselves. Even if we accept a metaphysics of laws, within-mechanism interactions are interactions between 'local' components.

Second, it is not at all easy to account for within-mechanism interactions in terms of a 'local' metaphysics. Energy transformations in biological systems obey the laws of thermodynamics. But it is very difficult to reconcile a power ontology with what it seems to be a global principle, such as the law of conservation of energy. This is something that friends of powers themselves have recognised (cf. Ellis 2001). So, contra Illari and Williamson, a focus on practice seems in fact to imply the opposite conclusion: global principles like the laws of thermodynamics are needed for accounting for within-mechanism interactions (e.g., as studied by bioenergetics; see Nelson et al. (2008, 489); but only a metaphysics in terms of laws seems to offer an adequate account of such global principles; so, a metaphysics of laws is required for a characterisation of the metaphysics of mechanisms. Again, the point here is that 'local' decompositions of mechanistic parts must be kept distinct from 'global' or 'local' ways to characterise interactions.

Third, there is a historical point to be made against the argument that mechanistic explanation is not compatible with a metaphysics of laws. This combination ('local' mechanisms that produce phenomena plus laws of nature) was a dominant view in seventeenth-century mechanical philosophy, as we saw in Chapter 1. Contemporary mechanistic explanations, of course, are very different from their seventeenth-century counterparts, which in many cases just involved parts of matter in motion. But the general pattern of explanation is similar: in giving a mechanistic explanation, one shows how the particular properties of the parts, their organisation and their interactions (which can be captured in terms of the laws that govern them) produce the phenomena.

In view of the previous points, premise (3) above can only be accepted if the meaning of 'local' is disambiguated. An option here is to say that mechanisms have to be local, in the sense that within-mechanism interactions have to be grounded in facts in the vicinity of the mechanism. So, one can think of causation as a local matter, that is, as a relation between the two events that are causally connected, and not as a global matter, that is, as involving a regularity. But note that so-called singular causation is compatible with a metaphysics of laws. One can view causation as a relation between 'local' events, but at the same time adopt an ontology of laws, where laws could be, for example, necessitating relations between universals, or Humean regularities, that is, 'global' facts about the universe (recall Chapter 5).

Note that Illari and Williamson themselves seem to recognise that in understanding scientific practice one need not talk about metaphysics, for they say: 'Understanding the metaphysics of mechanisms on this level is now a philosophical problem with no immediate bearing on scientific method, of course' (Illari & Williamson 2011, 834). But they add: 'It does, however, bear on our understanding of science' (p. 834). While we agree with the first sentence, we believe (and we shall argue below) that an understanding of mechanisms as causal pathways underpinned by difference-making relations is all one needs in order to understand scientific practice. We conclude, then, that there is no reason coming from scientific practice for accepting a power-based or an activities-based account of mechanism.

6.7 Against Glennan on Causation as Production

As we saw in Chapter 4, according to CM difference-making relations are enough to understand mechanisms and hence mechanistic causation and explanations. But those philosophers that view causation in terms of production contest this point. Glennan (2017) is one of the defenders of this view. According to him, mechanisms, qua productive, are the truth-makers of causal claims:

> (MC) A statement of the form 'Event c causes event e' will be true just in case there exists a mechanism by which c contributes to the production of e. (p. 156)

Actually, there are as many causal relations as there are activities. As he puts it: 'There is on this [New Mechanist] view no one thing which is interacting or causing, and when we characterise something as a cause, we are not attributing to it a particular role in a particular relation, but only saying that there is some productive mechanism, consisting of a variety of concrete activities and interactions among entities' (p. 148). This pluralist view leads him to the radical conclusion that '[t]here is ... no such thing as THE ontology or THE epistemology of THE causal relation, but only more localised accounts connected with the particular kinds of producing' (p. 33).

MC tallies with Glennan's singularism about causation. All causings are singular and in fact fully distinct from each other. Singularism is committed to the view that causation is internal (intrinsic, as Glennan puts it) to its relata. Glennan shares this intuition. He says: 'Productive causal relationships are singular and intrinsic. They involve continuity from cause to effect by means of causal processes' (p. 154).

But is causation a relation, after all? And if yes, what are the *relata*? 'Events' is the answer that springs to mind. Glennan agrees but takes events to involve activities: 'Events are particulars – happenings with definite locations and durations in space and time. They involve specific individuals engaging in particular activities and interactions' (p. 149). Or as he put it elsewhere: 'an event is just one or more entities engaging in an activity or interaction' (p. 177).

We have already argued in the previous sections that activity is far from being a sui generis ontic category. Besides, there is the received account of events as property-exemplifications: events are exemplifications of properties (or relations) by an object (or set of objects) at a time (or a period of time). As Glennan admits: 'If exemplifying a property were the same as engaging in an activity, then the two views would coincide.' However, he takes it that 'there are important differences between exemplifying properties and engaging in activities' (p. 177).

The chief difference between property-exemplification and engaging in activities is, Glennan says, that 'properties are paradigmatically synchronic states of an entity that belong to that entity for some time'. Unlike activities, properties 'do not involve change'. Events, Glennan argues, 'involve changes'. It is indeed true that events involve change. The collision of the *Titanic* with the iceberg took time and during it, both the *Titanic* and the iceberg suffered changes in their properties, which resulted in another event, namely the sinking of the *Titanic*. It is true that to account for this we have to introduce relations: the collision is between the *Titanic* and the iceberg. But relations, we are told, are not 'activity-like'. Glennan insists that 'only events (which involve activities) can be causally productive'. Properties, he says, 'cannot produce anything' (p. 178).

When all is said and done, the key question is: Is causation production? Or is it difference-making? Glennan is clear: 'While I grant that production and relevance are two different concepts of cause, I will argue that production is fundamental' (p. 156).

Descriptively, Glennan distinguishes between three kinds of productive relations:

- Constitutive production: An event produces changes in the entities that are engaging in the activities and interactions that constitute the event.
- Precipitating production: An event contributes to the production of a different event by bringing about changes to its entities that precipitate a new event.
- Chained production: An event contributes to the production of another event via a chain of precipitatively productive events. (p. 179)

All this is fine, but what is the chief argument for causation being *production*?

It seems to be this: 'Mechanisms provide the ontological grounding that allows causes to make a difference' (p. 165). Glennan's problem with the claim that a mechanism is itself a network of relations of difference-making between events is that on the difference-making account 'the causal claim depends upon the truth of a counterfactual, whereas on the mechanist account the truth depends upon the existence of an actual mechanism' (p. 167). Furthermore, it is claimed that the truth of the counterfactual requires contrasting an actual situation – where the cause occurs – and a non-actual but possible situation in which the cause does not occur.

Does the production account avoid counterfactuals? Glennan acknowledges that causation as production relies on some notion of relevance but takes this to require actual difference-makers. He takes it that actual difference-makers are 'features of the actual entities and their activities upon which outcome depends' (p. 203).

What is an *actual* difference-maker? A factor such that had it not happened, the effect would not have followed. But (1) in an actual concrete sequence of events which brought about an effect x, all events were necessary in the circumstances; all were difference-makers. If any of them were absent, the effect, in its full concrete individuality, would not follow. A different effect would have followed. But (2) what makes true the counterfactual that 'had x not actually happened, y would not have followed'? To 'delete' x from the actual sequence is to envisage a counterfactual sequence (i.e., a distinct sequence of events) without x. It is then to compare two sequences: the actual and the counterfactual. This requires thinking in terms of counterfactual difference-making. What makes the counterfactual true is not the actual sequence of events but the fact, if it is a fact, that xs are followed by ys, which is a causal law.

Take the example of a ball striking a window while a canary nearby sings. The actual causal situation – the mechanism in all its particularity – includes the process of the acoustic waves of the canary's singing striking the window (say, for convenience, at the moment when the ball strikes the window) as well as the kinetic energy of the ball (which was a red cricket ball) and so on. Despite the fact that the acoustic waves are part of the actual concrete mechanism and clearly contributed to the actual breaking (no matter how little), we would not say that it was the singing that caused the window-breaking. It clearly didn't make a substantial contribution to the breaking. Had it not been there, the window would still have shattered. How can *this* counterfactual be made true by the actual situation? In the actual situation, the singing was a difference-maker since it was part of the mechanism that made the difference. To show that it did not make a difference (better put, that it made a difference without a difference) we have to compare the actual situation in which the singing took place and a non-actual but possible situation in which the singing did not happen. Whatever makes this counterfactual true, it is not the actual situation, in and of itself.

In sum: not only does production not avoid counterfactuals (if actual difference-makers are to be shown that did not make a difference), but it seems that the very idea of production requires difference-making relations if the producer of change is nothing more specific than everything that happened before the effect took place.

6.8   Activities and the Language of Science

In this last section we want to discuss some more general points concerning the methodology of new mechanists that extract metaphysical conclusions from scientific practice. New Mechanists, beginning with MDC, typically stress the 'descriptive adequacy' of characterising mechanisms in terms of organised entities and activities; activity talk, in particular, is commonly taken to be descriptively superior over more standard philosophical talk in terms of the metaphysics of substances, properties, relations and laws. So, a general reason for preferring a metaphysics of activities, which we take to be the underlying rationale of new mechanists that subscribe to activities, is that only such a metaphysics can make sense of the kind of talk one finds in science.

This move, however, gives rise to a crucial and general question that can be posed to all attempts to extract substantive metaphysical conclusions from the kind of talk one finds in a particular discourse, be it everyday language or scientific discourse: Is language a good guide to ontology? More specifically, should we be committed to activities (just) because scientists use verbs and gerunds to describe how mechanisms work? There are well-known arguments, which we tend to accept, that it is not straightforward to read off one's ontology from how language is used. For example, it is not a straightforward manner to decide what is the correct quantum ontology, or whether we should accept the category of relations as a fundamental ontological category distinct from entities and (monadic) properties based simply on certain linguistic considerations. Frank Ramsey's (1925) well-known scepticism concerning the universal/particular distinction and whether we can read it off our language as well as Bertrand Russell's (1905) theory of definite descriptions illustrate some of the difficulties involved here. We have a similar approach concerning mechanical ontology: we think that there are no good reasons to read off a mechanical ontology of activities from the way that molecular biologists, for example, talk.

This thesis, that scientific discourse does not, on its own, favour a particular metaphysical account, may seem to be at odds with the kind of argumentation we used against Glennan's account of activities in Section 6.5. We argued there that activities are just manifestations of powers. One could wonder, then, whether by saying this we are committed to a metaphysics of properties and powers, which, if scientific discourse has no specific metaphysical implications, may seem equally questionable as a scientific ontology. Do we perhaps think that when scientists talk about activities they are in reality talking about the manifestations of powers and so are committed to the existence of powers?

We think that when scientists talk about activities they don't thereby take it that they are reducible to properties or powers; but we think it's also correct to say that scientists do not have a view (i.e., a systematic philosophical conception) of the possible relations between activities and more traditional metaphysical categories. Hence, in line with the general point made above about the relationship between language and ontology, scientists' talk of activities (such as bonding, transmitting, interrupting) does not imply that there really exist activities in the world as a fundamental ontological category. Of course, this is also true regarding other kind of ontologies; in thinking about mechanisms methodologically, and if we just focus on biologists' way of talking, we are not committed to, for example, a power-based ontology or a Humean one (i.e., categorical properties plus laws). For us this is exactly as it should be when it comes to the methodological role of mechanisms.[7]

But although we take both an activities-based metaphysics and other kinds of metaphysics (e.g., a metaphysics based on powers) to be not derivable from scientific practice, we also think that some metaphysical accounts may be conceptually incoherent or more inflated than others. In particular, we take an activities-based metaphysics to be such an inflated account (we also think that it is conceptually incoherent). We have argued in this chapter that according to standard mechanistic approaches, activities, if they are posited as a fundamental category, are taken as distinct and irreducible elements of reality. We do not take this to mean that activities can exist independently from entities and properties. We certainly agree with new mechanists that activities are not capable of existing without one or more entities engaging in them – the situation is similar to relations: relations require relata. Yet that's not the end of the story. Though activities depend on entities in that entities are required for activities, for some new mechanists, activities are irreducible to their entities and properties; they are a genuine ontological add-on. Hence an activities-based ontology is certainly more inflated ontology compared with an ontology of entities and properties.

To clarify, take again the example of relations. We may think of the God-metaphor and ask (as some metaphysicians have asked): Suppose that in the Creation God created all entities and their properties. Did he then rest? Or did he have to add relations? If he rested, then relations can be taken as internal (i.e., given entities with their properties, we thereby have all relations). If he didn't, then (at least some) relations are external; that is, they are something over and above entities and properties. These two broad ontological pictures have to be assessed on the basis of general metaphysical considerations, as well as on the basis of their compatibility with current science. The same is true for mechanical ontologies postulating activities: to take activities as fundamental is to buy into a more inflated metaphysics, where activities enter the world on top of entities and properties (just as in the case of relations). For example, using the God metaphor, the friends of activities would say that, when God created the

---

[7] There is a possible exception here. It is plausible that our commitment to causation as, ultimately, a difference-making relation of (counterfactual) dependence requires a prior commitment to laws as an indispensable part of the truth-makers of counterfactuals (see Chapter 7). But note that even at this high level of generality, there needn't be a commitment to a specific account of lawhood.

chemical elements and their properties (e.g., valence), he didn't thereby fix all facts about chemical compounds; instead, he had also to introduce the activity of 'chemical bonding' as a distinct ontological item.

The question for us, then, is: Does such an ontological picture of added-on activities make sense on general philosophical grounds? There may be reasons why some ontological positions may be better than others, reasons unrelated (or at least not directly related) to the way scientists talk; in this chapter, we have examined some of those reasons as far as activities and production-based mechanical ontologies are concerned. In particular, we think that activities-based ontologies are both inflated compared with more sparse ontologies and not conceptually coherent. Ontologies that include properties, powers and relations, but lack activities, while coherent conceptually, are in turn inflated compared with a Humean ontology. Last, note that all this is not to say that our own methodological position is committed to an ontology of properties (say, qua universals or classes of resembling tropes or what have you), instead of to an ontology of activities. What we argue is that there exist good independent reasons to be sceptical of an activities ontology.

CHAPTER 7

Whither Counterfactuals?

7.1 Preliminaries

There have been two traditions concerning how the 'link' between cause and effect is best understood (Hall 2004; Psillos 2004). According to the first tradition, which goes back to Aristotle, there is a *productive relation* between cause and effect: the cause produces, generates or brings about the effect. This productive relation between cause and effect has been typically understood in terms of powers, which in some sense ground the bringing-about of the effect by the cause. According to the second tradition, which goes back to Hume, the link is some kind of robust relation of dependence between what are taken to be distinct events. On this account, the chief characteristic of causes is that they are *difference-makers*: the occurrence of the cause makes a difference to the occurrence of the effect.[1]

We have already discussed in detail the currently most popular version of the production approach that cashes out the link between cause and effect by reference to mechanisms. When it comes to difference-making, there have been various ways to understand this notion of dependence. It may be nomological dependence (cause and effect fall under a law), counterfactual dependence (if the cause hadn't happened, the effect wouldn't have happened) or probabilistic dependence (the cause raises the probability of the effect). But, arguably, the core notion of difference-making is counterfactual, that is, based on contrary-to-fact hypotheticals. That is, a causal claim of the form 'A caused B' would be

---

[1] Though traditionally causation has been taken to be a single, unitary concept, there has been a tendency, as of late, to question this assumption. The case for there being *two* concepts of causation has been made, quite forcefully, by Ned Hall in his 2004 paper, where he distinguishes between causation as *dependence* and causation as *production*. Hall takes dependence to be *counterfactual* dependence, while he takes the concept of production (*c* produces *e*) as primitive.

understood as implying: if A hadn't happened, B wouldn't have happened either. It is in this sense that A *actually* makes a difference for B.[2]

A currently popular version of the dependence approach is Woodward's *interventionist counterfactual* account, which takes the relationship among some variables X and Y to be causal if, were an intervention to change the value of X appropriately, the relationship between X and Y would remain invariant *and* the value of Y would change.

We shall argue that the interventionist approach, despite its undeniable attractions, faces a couple of important problems. The first relates to what fixes the truth-conditions of counterfactuals, while the second has to do with the role of laws of nature in grounding counterfactuals. Hence, in the end we will favour a Lewis-style account. In any case, we will conclude with a role mechanistic information can play within a difference-making account of causation.

7.2 Counterfactuals: A Primer

7.2.1 The Logic

Subjunctive conditionals or counterfactual conditionals are probably as old as language itself since they give speakers the means to talk about what would or might happen or have happened if certain things were to happen or had happened. In ordinary language, they have the form:

If x were (not) the case, then y would (not) be the case.

or

If x had (not) been the case, then y would (not) have been the case.

Subjunctive conditionals leave open the possibility of the realisation of whatever is expressed in the antecedent, for example, if John were to come to the party, Mary would not go. Counterfactual (or 'contrary-to-fact') conditionals are such that the antecedent is false; the state of affairs expressed in it has not actually obtained, for example, if John had gone to the party, Mary would not have gone. (Here it is an implicit assumption that the actual course of events is that John did *not* go to the party.) Both

---

[2] On a nomological account of causal dependence (i.e., B depends on A if there is a law that connects the two), counterfactuals are required to account for the modal strength of laws (for more on this, see Psillos 2002). So, even if it were to be admitted that the alternative notions of dependence are distinct, counterfactuals play a key role in all versions of the dependence approach to causation.

kinds of conditional contrast to indicative conditionals of the form: *if x is the case, then y is the case*. Though there are differences between them, we will not be detained by them, and concentrate on counterfactual conditionals. (From now on, we will follow customary usage and use $\square\!\!\rightarrow$ to express the counterfactual 'if . . ., then . . .'.)

Counterfactuals fail a number of principles that indicative conditionals satisfy. Most importantly, they are non-monotonic; that is, they fail the principle of strengthening of the antecedent:

$$X\square\!\!\rightarrow Y \text{ does not entail } X\&Z\square\!\!\rightarrow Y.$$

Example: If John had been poisoned, he would have died. This does not entail: if John had been poisoned and taken an antidote, he would have died.

Transitivity:

$$X\square\!\!\rightarrow Y \text{ and } Y\square\!\!\rightarrow Z \text{ does not entail } X\square\!\!\rightarrow Z.$$

Example: If John had gone to the market, he would have taken the bus; if John had taken the bus, then he would have gone to his office. These two do not entail: if John had gone to the market, then he would have gone to his office.

Contraposition:

$$X\square\!\!\rightarrow Y \text{ does not entail } \text{not-}Y\square\!\!\rightarrow \text{not-}X.$$

Example: If John had lived in a euro-zone country, he would have used euros. This is not equivalent to: if John had not used euros, he would not have lived in a euro-zone country.

If we assume that classical semantics apply to indicative conditionals (the indicative conditional is true iff either the antecedent is false or the consequent is true), trying to apply classical semantics to counterfactuals leads to their trivialisation: given the actual falsity of the antecedent of a counterfactual, both the counterfactual with the actual consequent and the counterfactual with the negation of the actual consequent end up being true.

Example: given that the vase was not struck with a hammer, both of the following two conditionals (treated as material conditionals) are true:

If this vase had been struck with a hammer, it would have broken.

and

If this vase had been struck with the hammer, it would not have broken.

The failure of the three principles and this unwanted consequence is a *reductio* of the view that classical semantics apply to counterfactuals. But then, what is the right semantics for counterfactuals? What are the truth-conditions of a counterfactual conditional? Or, at least, what are their assertibility conditions? This problem came under sharp focus in the 1940s, when philosophers started to realise that the concept of counterfactual conditionals is instrumental for the explication/understanding of a number of other philosophical concepts. As Nelson Goodman put it in one of the first papers to deal with this issue: 'if we lack the means of interpreting counterfactual conditionals, we can hardly claim to have any adequate philosophy of science' (1947, 113).

Note that in assessing a counterfactual assertion X$\square\!\!\rightarrow$ Y, we should replace, as it were, the actual non-occurrence of X with the supposition that X has occurred. But given that the laws of nature and the actual course of events led to non-X, in supposing the actual occurrence of X we need to make counterfactual suppositions concerning either the laws or the actual course of events such that X actually occurred. In particular, we have to assume either that some laws were broken (so that X did happen after all) and/or that some actual particular matters of fact did not occur. Hence, in specifying the semantics of counterfactuals, we have to take into account considerations concerning the laws of nature and other particular matters of fact prior to the conditions specified in the antecedent of the counterfactuals.

There are two major views concerning the semantics of counterfactuals, the first being introduced by Goodman himself, while the second was developed by Robert Stalnaker and David Lewis (but introduced by William Todd in 1964). Let us examine them in turn.

7.2.2   The Semantics

7.2.2.1   The Metalinguistic or 'Support' View

On the first major view, known as 'support view' or 'metalinguistic view', a counterfactual conditional X$\square\!\!\rightarrow$ Y is an elliptic or telescoped argument (or a linguistic construction *about* an argument) such that the antecedent X (taken in its indicative form) together with suitable auxiliary premises entails the consequent (taken in its indicative form). Hence, X$\square\!\!\rightarrow$ Y should not be taken to be a statement at all; its assertoric content is captured by the following argument-type:

$$X \& S \& L \text{ (materially) imply } Y,$$

where L are statements capturing laws of nature and S are singular statements capturing background or collateral conditions which should be 'cotenable' with the antecedent X and express necessary conditions for the consequent to follow.

Example: If this match had been struck (X), it would have lit (Y). For a struck match to light, it is necessary that the match is well made, that it is dry, that there is oxygen and so on. But even these conditions (collectively designated by S) are not sufficient for the lighting of the match; various laws are required (collectively designated by L). Hence, in asserting the counterfactual 'if this match had been struck, it would have lit', we are committed to the truth of the various statements that describe the laws and the relevant background conditions.

The first general problem with this view concerns the characterisation of the relevance relation when it comes to the background/collateral conditions. It cannot be too permissive. If we allowed all true statements to be relevant to the argument, the falsity of the antecedent X (which is *actually* false) would be relevant too; but then the counterfactual would be trivially true. It cannot be too restrictive either. The consequent of the counterfactual is false too. Hence not-Y is the case. It's not hard to see that given $(X \& S \& L \rightarrow Y)$ and not-X and L, it follows (by obvious steps) that $X \rightarrow$ not-S. (Assuming, for simplicity, that S is 'the match is dry', the conclusion would be: if the match is struck, it will not be dry!) The point, then, is that only those background/collateral conditions which are 'cotenable' with the antecedent should be admitted. But which are they? Those conditions S which are such that if X had been true, S would have been true too. This is a counterfactual assertion, and Goodman thought that this kind of circularity impairs the metalinguistic analysis of counterfactuals.

The second general problem with this view concerns the characterisation of laws, which are indispensable for the *connection* between the antecedent and the consequent of a counterfactual. The key thought here is that some generalisations, though true, are unable to 'support' counterfactuals because they are accidental. *Example*: Compare the following two counterfactuals:

> (A)   If x had been a golden sphere, its diameter would not have been more than one mile long.
> (B)   If x had been a plutonium sphere, its diameter would not have been more than one mile long.

(A) is false, while (B) is true – we rightly suppose. And this is because there is a law of nature backing up (B), while the generalisation related to (A) is merely accidental (intuitively: if we have had enough gold, we could build a sphere of it with the required diameter; not so with plutonium). So the general statements L that are part of the premises of the argument whose telescopic form is $X \square \rightarrow Y$ must express *laws of nature* and not merely accidentally true generalisations. But how exactly are we to distinguish between laws and accidents? If we felt that laws are those generalisations that support counterfactuals while accidents are those that do not (see the example above), then we would move in a(nother) circle. So there is need to look for ways to distinguish between laws and accidents that do not rely (in the first instance, at least) on their modal force (at least when expressed in their support of counterfactuals). When Goodman brought this problem to the attention of philosophers, the prevailing view of laws was that they are simply regularities (cf. Chisholm 1946); hence the distinction between laws and accidents (which are regularities too) was taken to be mostly an 'honorific' distinction which is captured by the different epistemic attitudes we have towards them. For instance, laws are those regularities that are projected to the future or are conformable by their instances and so on. Wilfrid Sellars (1958, 268), however, pointed out that even if laws are taken to be regularities, they are those regularities that are characterised by 'neck-sticking-out-ness', where this characteristic is captured in the subjunctive mood: 'If this were an A-situation, it would be accompanied by a B-situation.' The counterfactual content of a law, then, is seen as a 'contextual implication' of a law-statement.

This idea of 'contextual implication' is captured by the supposition view of counterfactuals which is akin to (though interestingly different from) the metalinguistic view, as this was developed by John Mackie (1973). As we already saw in Section 5.3.1, according to this view, to assert something like $X \square \rightarrow Y$ is to assert Y *within the scope of the supposition that X*. In other words, we suppose X and then we envisage various possibilities and consequences. This account brings to light the contextuality of counterfactual conditionals, which is not resolvable without some degree of arbitrariness: X did not happen; supposing that X did happen, what else do we have to assume or suppose? What features of the background (including laws and particular matters of fact) should we retain or change? There is no uniquely determined answer to this question, though contextual matters (including a fuller specification of the antecedent of the conditional) might (and as a rule do) help us. *Example*: Is the counterfactual: 'If I had let go of this stone, it would have fallen to the ground' true or false (or assertible/not-assertible)? It depends on the context! There are certain conversational contexts, in which it would be false to assert it, for example, if this were a precious stone and the owner was very careful with it, so if the stone were to be let go, it would have been caught in mid-air. *Another example*: Consider the following pair of counterfactuals: 'If Julius

Caesar had been in charge of UN Forces during the Korean War, then he would have used nuclear weapons' and 'If Julius Caesar had been in charge of UN Forces during the Korean War, then he would have used catapults.' Only contextual assumptions can tell us which one, if any, and in what context is true (or assertible).

The supposition view takes it that counterfactuals are *not* truths about possible words but ways to express an attitude towards a possible state of affairs made within the scope of a supposition. Suppose that the sole ground for believing the law L (e.g., All Fs are G) is an enumeration of *all* actual instances (Fai and Gai) of L. Then adding the supposition X, namely, that a *further* a is F, removes the ground for accepting L. We can no longer draw the conclusion that this further a is G. Hence we cannot assert the counterfactual X□→ Y. More generally, if the reasons for accepting L survive placing L within the scope of *the supposition that there are further instances of the law's subject term*, then we can say that the law supports the relevant counterfactual conditional. The required reasons are ordinary inductive reasons, namely, good inductive evidence for the law. Good inductive evidence, in other words, is evidence for the 'neck-sticking-out-ness' of the law.

An interesting related thought comes from Julius Weinberg (1951), who claimed the following. A counterfactual X□→ Y is best seen not as the indicative statement (the statement of a generalisation) X → Y plus some further antecedent conditions (including that X did not actually happen) but rather as asserting something about the evidence there is for X → Y, namely, that there is evidence for, and no evidence against, the generalisation: for all X (X → Y). Hence, the additional strength a counterfactual is supposed to have over the corresponding generalisation is captured by the evidence there is for the generalisation.

7.2.2.2   *The Possible-Worlds View*

Taking literally the view that counterfactuals are used in contemplating *possibilities*, the second major view of the semantics of counterfactuals appeals to possible worlds. In first suggesting this view, Todd (1964, 107) noted that when we allow for the possibility that the antecedent of a counterfactual be true, we are 'hypothetically substituting a different world for the actual one'. On this view, the core meaning of a counterfactual X□→ Y is (roughly): In the possible (but not actual) world where X, Y too.

A possible world is a way the world might be or might have been. For instance, it is possible that gold is not yellow, that planets describe circular orbits, that birds do not fly or that beer doesn't need yeast to brew. But are there really possible worlds? There are three views here. The first is that talk of possible worlds is a mere *façon de parler*, though useful when it comes to assessing counterfactuals. (We take it that an extension of this view is that possible worlds are useful *fictions*.) The second is 'extreme realism', according to which the way the world *actually* is, is one among the many ways the world could be; hence, the actual world is one among the many possible worlds, the latter being no less real than the actual. The chief advocate of this view was David Lewis (1973). The third view is 'abstract realism', according to which possible worlds are maximally consistent sets of propositions: total ways things might be. A 'possible world' then is fit to represent a concrete reality, but only one possible world actually represents anything, namely, the actual world (see Bennett 2003).

Stalnaker (1968) developed the core meaning of counterfactuals as follows:

Consider a possible world W in which X is true, but is otherwise similar to the actual world @. X□→ Y is true iff Y is true in W.

The similarity relation among worlds (a selection function, as Stalnaker put it) is an ordering of possible worlds with respect to their resemblance to the actual world.

Calling an X-world a possible world in which X holds, counterfactuals might be taken to be strict conditionals of the form:

X□→ Y is true in a world W iff Y is true in all X-worlds such that ___ .

where the blank is filled by a general condition that X-worlds should satisfy. Hence, whatever goes into the blank places a restriction on the admissible (or accessible) possible worlds. This idea would model counterfactuals along the lines of strict conditionals of the form:

It is physically necessary that ___

or

It is logically necessary that ___

where the first restriction is to all worlds with the same laws as the actual, while the second 'restriction' would be to all possible worlds *simpliciter*.

But this analysis cannot be correct. There is no set of possible worlds W such that X → Y throughout W (this is another way to state the fact that counterfactuals are non-monotonic). So Lewis (1973) suggested that counterfactuals X□→ Y are *variably strict conditionals*: each of them is a strict conditional, i.e., every X-world *of a certain sort* is a Y-world; but the relevant set of worlds varies with different conditionals.

Like Stalnaker, Lewis takes it that worlds are ordered in terms of similarity, or closeness to the actual world. According to this *primitive* notion of 'comparative overall similarity': 'we may say that one world is closer to actuality than another if the first resembles our actual world more than the second does, taking account of all the respects of similarity and difference and balancing them off against one another' (1986b, 163).

But unlike Stalnaker, Lewis took it that in assessing the counterfactual $X \square\rightarrow Y$ it does not make good sense to talk about *the* closest-to-actual possible X-world. It's not just that there might be more than one closest-to-the-actual possible worlds. It is mainly that there might not be even one rightly deemed *the* closest (even in a limiting sense). Hence, on Lewis's view:

> $X \square\rightarrow Y$ is true at a world W iff some (accessible) X-world in which Y holds is closer to W than any X-worlds which Y does not hold.

For instance, take the counterfactual that if this pen had been left unsupported (X), it would have fallen to the floor (Y). Neither X nor Y is true of the actual world. The pen was never removed from the table, and it didn't fall to the floor. Take all X-worlds. The counterfactual $X \square\rightarrow Y$ is true (in @) iff the X-worlds in which Y is true (i.e., the pen is left unsupported and falls to the floor) are closer to @ than any of the X-worlds in which Y is false (i.e., the pen is left unsupported but does not fall to the ground, e.g., it stays still in mid-air). As Lewis (1986b, 164) put it: 'a counterfactual ... is true iff it takes less of a departure from actuality to make the consequent true along with the antecedent than it does to make the antecedent true without the consequent'.

The key idea behind the possible-world semantics is that in specifying the truth-conditions of a counterfactual conditional we should imagine a state of affairs in which X obtains and which is such that *all else is pretty much as they actually were.* But as noted already, this is not quite possible. In the possible world in which X did happen, many other things (including the laws) were different from the actual world @ in which X did not occur. Can we find comfort in the notion of comparative similarity? Now, though 'comparative overall similarity' is not strictly defined, a lot can be said of it. Notably, it imposes a weak ordering on the set of possible worlds which are accessible from @, namely, the relation of comparative similarity is connected and transitive. (It also imposes a centring assumption: @ is closer to itself than any other world is to it.) More importantly, however, similarity is clearly not one-dimensional, but rather the resultant of many component similarities. Lewis (1986a, 47–8) ranks possible worlds according to the following dimensions of similarity (put in order of importance).

- Avoid big, widespread violations of the laws of nature of the actual world (very important).
- Maximise the spatiotemporal perfect match of particular matters of fact.
- Avoid small, localised violations of the laws of nature of the actual world.
- Secure approximate similarity of particular matters of fact (not at all important).

So, a world W_1 which has the same laws of nature as the actual world @ is closer to @ than a world W_2 which has different laws. But insofar as there is exact similarity of particular facts in large spatiotemporal regions between @ and a world W_3, Lewis allows that W_3 is close to @ even if some of the laws that hold in @ are violated in W_3.

All this implies that there is quite a lot of vagueness in the notion of overall comparative similarity, which accounts for the fact that counterfactuals themselves are vague, at least in the sense that it is a contextual matter what to keep fixed and what to change when we assert a counterfactual conditional. A more serious worry relates to the issue of the motivation behind the foregoing ranking of dimensions of similarity among worlds. It has been observed by many that Lewis's initial theory yielded the wrong truth-values for a type of counterfactual conditional which can be schematised thus:

$$X \square\!\!\rightarrow \text{BIG DIFFERENCE.}$$

For instance:

(C) If the president had pressed the button, a nuclear war would have ensued.

Intuitively, (C) is true. But on Lewis's initial account, it would be false. For a possible world W_1 in which the president did press the button and a nuclear war did erupt is more distant from (because more dissimilar to) the actual world than a world W_2 in which the president did press the button but, somehow, a nuclear war did *not* follow. Addressing this worry, Lewis noted that intuitive judgements of the truth and falsity of counterfactuals are prior to the similarity relation that is required for the semantics of counterfactuals; hence the similarity relation should be such that it tallies with the right intuitive judgements concerning counterfactuals. The similarity ranking above is meant to solve this problem. To see how the foregoing counterfactual is indeed true, Lewis invites us to consider the following. Take a world W_1 in which nothing extraordinary happened between the president's pressing the button and the activation of the nuclear missiles. In W_1 the nuclear war did erupt. Take, now, a world W_2 in which the president did press the button but the nuclear war did *not* follow. For this to happen, many miracles would need to take place (or, to put it in a different way, a really *big* miracle would have to occur). For all the many and tiny traces of the button pushing would have to be wiped out. Hence, appearances to the contrary, W_2 would be more distant from (because more dissimilar to) actuality @ than W_1. The *big* violation of laws of nature in W_2 is outweighed by the maximisation of the perfect spatio-temporal match of particular matters of fact between W_1 and @. So, with the help of the refined criteria of similarity among possible worlds, the president-counterfactual comes out true. Still, one may follow Horwich (1987, 171–2) in wondering how psychologically plausible Lewis's theory becomes: the similarity criteria are so tailored that the right counterfactuals come out true, but they have little to do with our pre-theoretical under-standing of judgements of similarity.

As noted already in relation to the 'support' view, there are two hurdles that an adequate theory of counterfactuals has to jump. The first relates to cotenability. Lewis solves this problem by taking it that some conditions S are cotenable with X (the antecedent of the counterfactual X$\square\!\!\rightarrow$ Y) iff some X-world is closer to the actual world than any not-S world. The second hurdle relates to the distinction between laws and accidents. Here, the possible-world approach is on safe ground, though the ground can support any decent theory of counterfactuals. David Lewis (1973) revamped a long tradition that goes back to John Stuart Mill, via Frank Ramsey, according to which the regularities that constitute the laws of nature are those that are expressed by the axioms and theorems of an ideal deductive system of our knowledge of the world and, in particular, of a deductive system that strikes the *best* balance between simplicity and strength. Simplicity is required because it disallows extraneous elements from the system of laws. Strength is required because the deductive system should be as informative as possible about the laws that hold in the world. Whatever regularity is not part of this *best system* is merely accidental. The gist of this approach is that no regularity, taken in isolation, can be deemed a law of nature. The regularities that constitute laws of nature are deter-mined in a kind of holistic fashion by being parts of a structure. An advantage of this approach is that it can sustain, in a non-circular way, the view that laws can support counterfactuals. For it identifies laws *independently* of their ability to support counterfactuals.

A key objection to the possible-world approach to counterfactuals is that counterfactual conditionals are not purely objective; an irremediably subjective element enters into the judgement of similarity (and, arguably, into the distinction between laws and accidents). Not only are the truth-conditions of counterfactuals 'a highly volatile matter' as Lewis (1973, 92) himself noted, but also what counterfactuals are true turns out to depend on various partly non-objective judgements concerning similarity weights and conversational contexts. This objection, however, might not be as fatal as it first seems precisely because counterfactual conditionals should be taken not to be pointers to necessary connections, powers and the like, but (in either of the two theories we have examined) summaries of attitudes we have towards statements that are supposed to express a *connection* between a hypothetical antecedent and a consequent. To exploit an idea of Sellars's, the core idea behind counterfactual reasoning is to assert that there are not good inductive reasons to affirm simultaneously a generalisation and the physical possibility of an exception to it.

7.3 Counterfactual Manipulation and Causation

In a series of papers and a book, Woodward (1997; 2000; 2003a; 2003b) developed a new account of the semantics of counterfactuals which he put to the service of a counterfactual account of causation. His account is *counterfactual* in the following sense: what matters is what *would* happen to a relationship, *were* interventions to be carried out. A relationship among some variables X and Y is causal if, were one to intervene to change the value of X appropriately, the relationship between X and Y wouldn't change *and* the value of Y would change. To use a stock example, the force exerted on a spring *causes* a change of its length, because if an intervention changed the force exerted on the spring, the length of the spring would change too (but the relationship between the two magnitudes – expressed by Hooke's law – would remain invariant, within a certain range of interventions).

Woodward (1997; 2000; 2003a) has analysed further the central notions of invariance and intervention. The gist of his characterisation of an *intervention* is this. A change of the value of X counts as an intervention I if it has the following characteristics:

1. The change of the value of X is entirely due to the intervention I.
2. The intervention changes the value of Y, if at all, only through changing the value of X.

The first characteristic makes sure that the change of X does not have causes other than the intervention I, while the second makes sure that the change of Y does not have causes other than the change of X (and its possible effects).[3] These characteristics are meant to ensure that Y-changes are exclusively due to X-changes, which, in turn, are exclusively due to the intervention I. As Woodward notes, there is a close link between intervention and manipulation. Yet his account makes no special reference to human beings and their (manipulative) activities. Insofar as a process has the right characteristics, it counts as an intervention. So, interventions can occur 'naturally', even if they can be highlighted by reference to 'an idealised experimental manipulation' (2000, 199).

Woodward links the notion of intervention with the notion of *invariance*. A certain relation (or a generalisation) is invariant, Woodward says, 'if it would continue to hold – would remain stable or unchanged – as various other conditions change' (p. 205). What really matters for the characterisation of invariance is that the generalisation remains stable under a set of actual and counterfactual *interventions*. So Woodward (p. 235) notes:

> the notion of invariance is obviously a modal or counterfactual notion [since it has to do] with whether a relationship would remain stable if, perhaps contrary to actual fact, certain changes or interventions were to occur.

Counterfactuals have been reprimanded on the ground that they are context-dependent and vague. Take, for instance, the following counterfactual: 'If the smoker had not smoked so heavily, they would have lived a few years more.' What is it for it to be true? Any attempt to say whether it is true, were it to be possible at all, would require specifying what else should be held fixed. For instance, other aspects of the smoker's health should be held fixed, assuming that other factors (e.g., a weak heart) wouldn't cause a premature death anyway. But what things to hold fixed is not, necessarily, an objective matter.

7.3.1   *Interventions*

To address this problem, Woodward devises a new theory of the semantics of counterfactuals. The key idea is that only counterfactuals which are

---

[3] There is a *third* characteristic too, namely, that the intervention I is not correlated with other causes of Y besides X.

related to *interventions* can be of help when it comes to assessing their test- or assertibility conditions. An intervention gives rise to an 'active counter- factual', that is, to a counterfactual whose antecedent is made true by (hypothetical) interventions. Woodward (2003b, 3) very explicitly char- acterises the appropriate counterfactuals in terms of *experiments*: they 'are understood as claims about what would happen if a certain sort of experiment were to be performed'.

Consider a case he discusses (pp. 4–5). Take Ohm's law (that the voltage E of a current is equal to the product of its intensity I times the resistance R of the wire) and consider the following two counterfactuals:

(1) If the resistance were set to $R = r$ at time t, and the voltage were set to $E = e$ at t, then the intensity I would be $i = e/r$ at t.

(2) If the resistance were set to $R = r$ at time t, and the voltage were set to $E = e$ at time t, then the intensity I would be $i* \neq e/r$ at t.

There is nothing mysterious here, says Woodward, 'as long as we can describe how to test them' (p. 6). We can perform the experiments at a future time $t*$ in order to see whether (1) or (2) is true. If, on the other hand, we are interested in what *would* have happened had we performed the experiment in a past time t, Woodward invites us to rely on the 'very good evidence' we have 'that the behaviour of the circuit is stable over time' (p. 5). Given this evidence, we can assume, in effect, that the *actual* performance of the experiment at a future time $t*$ is as good for the assessment of (1) and (2) as a *hypothetical* performance of the experiment at the past time t.

For Woodward, the truth-conditions of counterfactual statements (and their truth-values) are not specified by means of an abstract metaphysical theory, for example, by means of abstract relations of similarity among possible worlds. He calls his own approach 'pragmatic'. That's how he puts it:

> For it to be legitimate to use counterfactuals for these goals [understanding causal claims and problems of causal inference], I think that it is enough that (a) they be useful in solving problems, clarifying concepts, and facil- itating inference, that (b) we be able to explain how the kinds of counter- factual claims we are using can be tested or how empirical evidence can be brought to bear on them, and (c) we have some system for representing counterfactual claims that allows us to reason with them and draw infer- ences in a way that is precise, truth-preserving and so on. (p. 4)

At the same time, Woodward's view is meant to be realist and objec- tivist. He is quite clear that counterfactual conditionals have non-trivial truth-values independently of the actual and hypothetical experiments by virtue of which it can be assessed whether they are true or false. He says:

> On the face of things, doing the experiment corresponding to the antecedent of (1) and (2) doesn't *make* (1) and (2) have the truth values they do. Instead the experiments look like ways of *finding out* what the truth values of (1) and (2) were all along. On this view of the matter, (1) and (2) have non-trivial truth values – one is true and the other false – even if we don't do the experiments of realizing their antecedents. Of course, we may not *know* which of (1) and (2) is true and which false if we don't do these experiments and don't have evidence from some other source, but this does not mean that (1) and (2) both have the same truth-value. (p. 5)

So though 'pragmatic', Woodward's theory is also objectivist. But it is minimally so. As he notes, his view

> requires only that there be facts of the matter, independent of facts about human abilities and psychology, about which counterfactual claims about the outcome of hypothetical experiments are true or false and about whether a correlation between C and E reflects a causal relationship between C and E or not. Beyond this, it commits us to no particular metaphysical picture of the 'truth-makers' for causal claims. (2003a, 121–2)

7.3.2   Truth-Conditions

There are a few delicate issues here to be reckoned with. We will restrict ourselves to the following: *What are the truth-conditions of counterfactual assertions?* Woodward doesn't take all counterfactuals to be meaningful and truth-valuable. As we have seen (see also 2003a, 122), he takes only a subclass of them, the active counterfactuals, to be such. However, he does not want to say that the truth-conditions of active counterfactuals are fully specified by (are reduced to) actual and hypothetical experiments. If he said this, he could no longer say that active counterfactuals have determinate truth-conditions independently of the (actual and hypothetical) experiments that can test them. In other words, Woodward wants to distinguish between the truth-conditions of counterfactuals and their evidence- (or test) conditions, which are captured by certain actual and hypothetical experiments. The problem that arises is the following. Though we are given a relatively detailed account of the evidence-conditions of counterfactuals, we are not given anything remotely like this for their *truth-conditions*. What, in other words, is it that makes a certain counterfactual conditional true?

A thought here might be that there is no need to say anything more about the truth-conditions of counterfactuals other than offering a Tarski-style meta-linguistic account of them of the form

(T) 'If *x* had been the case, then *y* would have been the case' is true iff if *x* had been the case, then *y* would have been the case.

This move is possible indeed but not terribly informative. We don't know when to assert (or hold true) the right hand-side. And the question is precisely this: When is it right to assert (or hold true) the right-hand side? Suppose we were to tell a story in terms of actual and hypothetical experiments that realise the antecedent of the right-hand side of (T). The obvious problem with this move is that the truth-conditions of the counterfactual conditional would be specified in terms of its evidence-conditions, which is exactly what Woodward wants to block. Besides, if we just stayed with (T) above, without any further explication of its right-hand side, *any* counterfactual assertion (and not just the active counter-factuals) would end up meaningful and truth-valuable. Here again, Woodward's project would be undermined. Woodward is adamant: 'Just as non counterfactual claims (e.g., about the past, the future, or unobserv-ables) about which we have no evidence can nonetheless possess non-trivial truth-values, so also for counterfactuals' (2003b, 5). This is fine. But in the case of claims about the past or about unobservables, there are well-known stories to be told as to what the difference is between truth- and evidence-conditions. When it comes to Woodward's counterfactuals, we are *not* told such a story.

In light of the above, there are two options available. The first is to *collapse* the truth-conditions of counterfactuals to their evidence-conditions. One can see the prima facie attraction of this move. Since evidence-conditions are specified in terms of actual and hypothetical experiments, the right sort of counterfactuals (the active counterfactuals) and only those end up being meaningful and truth-valuable. But there is an important drawback. Recall counterfactual assertion (1) above. On the option presently considered, what makes (1) true is that its evidence-conditions obtain. Under this option, counterfactual conditionals lose, so to speak, their counterfactuality. (1) becomes a shorthand for a future prediction and/or the evidence that supports the relevant law. If *t* is a *future* time, (1) gives way to an actual conditional (a prediction). If *t* is a past time, then, given that there is good evidence for Ohm's law, all that (1) asserts under the present option is that there has been good evidence for the law.

7.3.3 Laws to the Rescue?

In any case, Woodward seems keen to keep evidence- and truth-conditions apart. Then (and this is the *second* option available), some informative story should be told as to what the truth-conditions of counterfactual conditionals *are* and *how* they are connected with their evidence-conditions (i.e., with actual and hypothetical experiments). There may be a number of stories to be told here.[4] The one we favour ties the truth-conditions of counterfactual assertions to *laws of nature*. It is then easy to see how the evidence-conditions (i.e., actual and hypothetical experiments) are connected with the truth-conditions of a counterfactual: actual and hypothetical experiments are symptoms for the presence of a law. There is a hurdle to be jumped, however. It is notorious that many attempts to distinguish between genuine laws of nature and accidentally true generalisations rely on the claim that laws do, while accidents do not, support counterfactuals. So counterfactuals are called for to distinguish laws from accidents. If at the same time laws are called for to tell when a counterfactual is true, we go around in circles. Fortunately, there is the Mill-Ramsey-Lewis view of laws (see Psillos 2002, chapter 5). Laws are those regularities which are members of a coherent system of regularities, in particular, a system which can be represented as an ideal deductive axiomatic system striking a good balance between *simplicity* and *strength*. On this view, laws are identified independently of their ability to support counterfactuals. Hence, they can be used to specify the conditions under which a counterfactual is true.[5]

Let us consider here one relevant thought that is central to Woodward's approach. He takes laws to be relations that remain invariant under

---

[4] One might try to keep truth- and evidence-conditions apart by saying that counterfactual assertions have excess content over their evidence-conditions in the way in which statements about the past have excess content over their (present) evidence-conditions. Take the view (roughly Dummett's) that statements about the past are meaningful and true insofar as they are verifiable (i.e., their truth can be known). This view may legitimately distinguish between the *content* of a statement about the past and the present or future evidence there is for it. Plausibly, this excess content of a past statement may be cast in terms of counterfactuals: a meaningful past statement *p* implies counterfactuals of the form 'if *x* were present at time *t*, *x* would verify that *p*'. This move presupposes that there are meaningful and true counterfactual assertions. But note that a similar story *cannot* be told about counterfactual conditionals. If we were to treat their supposed excess content in the way we just treated the excess content of past statements, we would be involved in an obvious regress: we would need counterfactuals to account for the excess content of counterfactuals.

[5] Obviously, the same holds for the Armstrong-Dretske-Tooley view of laws (see Psillos 2002, chapter 6). If one takes laws as necessitating relations among properties, then one can explain why laws support counterfactuals and, at the same time, identify laws *independently* of this support.

(a range of) actual and counterfactual interventions. If this is so, when checking whether a generalisation or any relationship among magnitudes or variables is invariant we need to subject it to some variations/changes/interventions. What changes will it be subjected to? The obvious answer is: those that are permitted or are permissible by the prevailing laws of nature. Suppose that we test Ohm's law. Suppose also that one of the interventions envisaged was to see whether it would remain invariant if the measurement of the intensity of the current was made on a spaceship, which moved faster than light. This, of course, cannot be done, because it is a *law* that nothing travels faster than light. So, some *laws* must be in place before, based on considerations of invariance, it is established that some generalisation is invariant under some interventions. Hence, Woodward's notion of 'invariance under interventions' (2000, 206) cannot offer an adequate analysis of lawhood, since laws are required to determine what interventions are possible.

Couldn't Woodward say that even basic laws – those that determine what interventions and changes are possible – express just relations of invariance? Take, once more, the law that nothing travels faster than light. Can the fact that it is a law be the result of subjecting it to interventions and changes? Hardly. For it itself establishes the *limits* of possible interventions and control.[6] We do not doubt that it may well be the case that genuine laws express relations of invariance. But this is not the issue. For the manifestation of invariance might well be the *symptom* of a law, without being constitutive of it.[7]

Before we move on we want to address a possible objection. It might be that Woodward aims only to provide a *criterion* of meaningfulness for counterfactual conditionals without also specifying their truth-conditions. This would seem in order with his 'pragmatic' account of counterfactuals, since it would offer a criterion of meaningfulness and a description of the 'evidence conditions' of counterfactuals, which are presumed to be enough to understand causation. In response to this, we would not deny that Woodward has indeed offered a sufficient condition of meaningfulness. Saying that counterfactuals are meaningful if they can be interpreted as

---

[6] Woodward (2000, 206–7) too agrees that this law cannot be accounted for in terms of invariance.
[7] We take to heart Marc Lange's (2000) important diagnosis: either *all* laws, taken as a whole, form an invariant-under-interventions set, or, strictly speaking, no law, taken in isolation, is invariant-under-interventions. This does not yet tell us what laws *are*. But it does tell us what marks them off from intuitively accidental generalisations.

claims about actual and hypothetical experiments is fine (and a step forward in the relevant debate). But can this also be taken as a necessary condition? Can we say that *only* those counterfactuals are meaningful which can be seen as claims for actual and hypothetical experiments? If we did say this, we would rule out as meaningless a number of counterfactuals that philosophers have played with over the years. Consider the following pair of counterfactuals: 'If Julius Caesar had been in charge of UN forces during the Korean War, then he would have used nuclear weapons' and 'If Julius Caesar had been in charge of UN forces during the Korean War, then he would have used catapults.' It is hard to see how we could possibly tell which of them, if any, is true. And yet it's even harder to think of them as *meaningless*.

Take one of Lewis's examples, that had he walked on water, he would not have been wet. We don't think it is meaningless. One may well wonder what the point of offering such counterfactuals might be. But whatever it is, they are understood and, perhaps, are true. Perhaps, as Woodward (2003a, 151) says, the antecedents of such counterfactuals are 'unmanipulable for conceptual reasons'. But if they are understood (and if they are true), this would be enough of an argument *against* the view that manipulability offers a necessary condition for meaningfulness.

It turns out, however, that there are more sensible counterfactuals that fail Woodward's criterion. Some of them are discussed by Woodward himself (2003a, 127–33). Consider the true causal claim: Changes in the position of the moon with respect to the earth and corresponding changes in the gravitational attraction exerted by the moon on the earth's surface cause changes in the motion of the tides. As Woodward adamantly admits, this claim cannot be said to be true on the basis of interventionist (experimental) counterfactuals, simply because realising the antecedent of the relevant counterfactual is physically impossible. His response to this is an alternative way of assessing counterfactuals. This is that counterfactuals can be meaningful if there is some 'basis for assessing the truth of counterfactual claims concerning what would happen if various interventions were to occur'. Then, he adds, 'it doesn't matter that it may not be physically possible for those interventions to occur' (p. 130). And he sums it up by saying that 'an intervention on X with respect to Y will be "possible" as long as it is logically or conceptually possible for a process meeting the conditions for an intervention on X with respect to Y to occur' (p. 132). Our worry then is this. We now have a much more liberal criterion of meaningfulness at play, and it is not clear, to say the least, which counterfactuals end up meaningless by applying it.

7.4 Causal Inference and Counterfactuals

In the last decades, there has been an increasing interest in causal inference among statisticians and social scientists, and counterfactuals have loomed large in some key attempts to model it. Prominent among them is Rubin's model, which has been advanced by Donald Rubin (1978) and Paul Holland (1986).[8] This model focuses on the discovery of the effects of causes. Suppose, to use a simple example, we want to find out whether taking an aspirin makes a difference to a *specific* subject's relief from headache. We would like to give a certain subject u an aspirin in order to see what happens to the headache episode – let's call the result Y. But we would also like, at the same time, to withhold giving aspirin to the very same subject u, in order to see what happens to the headache episode – let's call this result Y'. The difference, if any, between Y and Y' would naturally be considered the actual causal effect of aspirin-taking on the headache episode of subject u. But this kind of experiment is impossible: the experimenter cannot both give and *not* give an aspirin to the *same* subject u at the *same* time. Rubin's and Holland's main idea is that an appeal to counterfactuals allows us to make an inference about the causal effect.

Let's consider a population U of individuals, or units, $u \in U$. In a typical experiment, the experimenter applies one treatment, say, i, out of a set of possible treatments, T, to each unit u and observes the resulting responses Y. The experimental units are chosen and separated into two groups (the experimental group and the control group) by randomisation. To simplify matters, let the treatment set T consist of two possible actions (treatment t, and control c). For instance, t may be taking the aspirin and c may be taking a placebo. Let, also, Y consist of two possible responses, for example, headache relief Y_t, and headache persistence Y_c. Though it is crucial that each unit is potentially exposable to any one of the treatments, to each unit u just one treatment is *actually* given, that is, either t or c. Similarly, for each unit u, there is just one response that is actually observed, that is, either $Y_t(u) = Y(t, u)$ or $Y_c(u) = Y(c, u)$. Rubin's model defines the two responses in counterfactual terms. That is, $Y(t, u)$ is the value of the response that would be observed if the unit u were exposed to treatment t and $Y(c, u)$ is the value that would be observed *on the same unit* u if it were exposed to c. A key assumption of Rubin's model is that both

---

[8] See also Holland (1988), Stone (1993), Cox and Wermuth (2001), Maldonado and Greenland (2002) and Kluve (2004).

values $Y(t, u)$ are $Y(c, u)$ are well defined and determined. In particular, it is assumed that even if subject u is actually given treatment t and has response $Y(t, u)$, there is still a fact of the matter about what the subject's u response would have been, had they been given treatment c. The task is to figure out the so-called *individual causal effect*, that is, the difference

(1) $\tau(u) = Y(t, u) - Y(c, u)$

which measures the effect of treatment t on u, relative to treatment c.

In each particular experiment, either $Y(t, u)$ or $Y(c, u)$ (but not both) ceases to be counterfactual. Yet, given that one of $Y(t, u)$ and $Y(c, u)$ becomes observable, the other *has to* be unobservable. Holland has called a situation such as this 'the *fundamental problem of causal inference*'. As he (1986, 947) put it: 'It is impossible to *observe* the value of $Y(t, u)$ and $Y(c, u)$ on the same unit and, therefore, it is impossible to observe the effect of t on u.' Does it follow that figuring out (1) above is impossible?

Suppose that we give treatment t to u and we observe $Y(t, u)$. The question, then, is how could we possibly figure out the value of $Y(c, u)$? Recall that $Y(c, u)$ is a counterfactual: the response that would be observed if the unit u were exposed to treatment c (given that it was in fact exposed to treatment t and the observed value was $Y(t, u)$). The important insight of Rubin's model is that when certain assumptions are in place, there are ways to assess counterfactuals such as the above. The following is how we may proceed.

Given that unit u got treatment t, we may try treatment c to a different unit u', which is very much like u, except that it was given treatment c instead. That is, instead of assessing the counterfactual conditional $Y(c, u)$, which is impossible, we assess the factual conditional $Y(c, u')$ – the response of unit u' if the subject is given treatment c – and claim that this tells *indirectly* what the value of $Y(c, u)$ is. If this move is to be plausible at all, we need an assumption of *unit homogeneity*. We need to assume that u and u' are so similar that the actual response of u' to treatment c is the same as the response that unit u *would* have to treatment c. Under this assumption, we take it that $Y(t, u) = Y(t, u')$ and $Y(c, u) = Y(c, u')$. Then, the individual causal effect can be calculated, since (1) becomes thus:

(2) $\tau(u) = Y(t, u) - Y(c, u) = Y(t, u) - Y(c, u')$.

This is all fine and we are prepared to say that, modulo the uniformity assumption, it does tell us something about the individual causal effect. But something strange has happened. Expression (1) involves essentially a counterfactual conditional $(Y(c, u))$; (2) does not. Expression (2) is indeed measurable, but the counterfactuals are gone. Instead, (2) has two factual conditionals, one for unit u who received treatment t and another for unit u' who received treatment c. In a sense, we are still asking: What would have happened to u, had we given it treatment c? But it also seems that we have now *reduced* this question to two different ones: (a) What *does* happen to u', if we give it treatment c? and (b) Assuming unit homogeneity, $Y(c, u')$ and $Y(t, u)$, what *is* the causal effect of t on u? These questions involve no counterfactuals. The content of the counterfactual conditional $Y(c, u)$ seems exhausted by the joined content of the factual conditional $Y(c, u')$ and the unit homogeneity assumption. In other words, the unit homogeneity assumption renders the counterfactual conditional $Y(c, u)$ not so much a claim about the *specific* unit u but rather a claim about *any* of the homogeneous units. It is because of this fact that the counterfactual becomes testable.

There is another way we might proceed in our attempt to calculate $\tau(u)$. This time, instead of giving treatment t to unit u and treatment c to (uniform) unit u', we give treatment c to unit u at time t_1 and treatment t to the *very same unit u* at a later time t_2. As Holland (1986, 948) notes, this move requires another assumption, namely, *temporal stability*. This, he says, 'asserts the constancy of response over time'. It also requires an assumption of 'causal transience', since it implies that 'the effect of the cause c and the measurement process that results in $Y(c, u)$ is transient and does not change u enough to affect $Y(t, u)$ measured later' (p. 948). So, if Alice's taking a placebo at time t_1 changes some properties of Alice enough to affect her response to taking an aspirin at a later time t_2, the causal effect of taking aspirin on Alice's headache episode ceases to be calculable. Under these assumptions, we take it that $Y(t_{t1}, u) = Y(t_{t2}, u)$ and $Y(c_{t1}, u) = Y(c_{t2}, u)$. If this is so, then the individual causal effect can be calculated, since (3) becomes thus:

$$(3) \quad \tau(u) = Y(t, u) - Y(c, u) = Y(t_{t2}u) - Y(c_{t1}, u).$$

The points made about (2) can be repeated about (3) too. Expression (3) has no counterfactuals and it seems that the content of (1) – which does involve the counterfactual $Y(c, u)$ – *reduces* to the joined content of two factual conditionals ($Y(t_{t2}, u)$ and $Y(c_{t1}, u)$) together with the two further assumptions of causal transience and temporal stability.

We are willing to allow that we may be wrong here. That is, it might be the case that counterfactuals such as the ones we have been discussing *do* have excess content over the joint content of the relevant factual conditionals and the relevant assumptions. Still, what matters is that counterfactual conditionals can be assessed in terms of truth and falsity only when certain assumptions are in place. Those assumptions might fail. If, however, there are reasons to believe they do not, then causal inference seems quite safe. This is really an important achievement of Rubin's model. But we shouldn't lose sight of the fact that these assumptions are characteristics of *stable causal or nomological structures*.[9] Consider *unit homogeneity*. For it to hold, it must be the case that two units u and u' are alike in all causally relevant respects other than treatment status. If this is so, we can substitute u for u' and vice versa. This simply means that there is a causal law connecting the treatment and its characteristic effect which holds for all homogeneous units and hence is independent of the actual unit chosen (or could have been chosen) to test it. In effect, this holds for temporal stability too, since the latter is the temporal version of unit homogeneity. It does indeed make sense to wonder what the value of the voltage in a resistor would have been if the intensity of the current was I instead of the actual I_0 precisely because Ohm's law provides a stable nomological structure to address this counterfactual. But suppose we wanted to check the counterfactual that had the election taken place at an earlier time, the government would have been re-elected. Here it is obvious that temporal stability cannot be assumed because there is no stable nomological structure to back it up. Law-backed counterfactuals can indeed be assessed precisely because the laws make sure that the required assumptions are in place.[10]

In light of the above, it might not be surprising that according to Judea Pearl (2000, 428), who is one of the champions of the counterfactual approach, 'the word "counterfactual" is a misnomer'. In the case of individual causal effects, Pearl notes, we are interested in finding out things such as this:

> Q_{II}: The probability that John's headache would have stayed had he not taken aspirin, given that he did in fact take aspirin and the headache has gone.

It does not matter for present purposes that Pearl formulates the issue in terms of probabilities. What matters is that Q_{II} is a counterfactual claim of which Pearl stresses:

---

[9] Our favourite way to spell out this notion is given by Herbert Simon and Nicholas Rescher (1966). In fact, in showing how a stable structure can make some counterfactuals true, they blend the causal and the nomological in a fine way.

[10] It goes without saying that this causal or nomological structure should be characterised independently of counterfactuals, but as Simon and Rescher (1966) show, this can be done.

Counterfactual claims are merely conversational shorthand for scientific predictions. Hence Q_{II} stands for the probability that a person will benefit from taking aspirin in the next headache episode, given that aspirin proved effective for that person in the past.... Therefore, Q_{II} is testable in sequential experiments where subjects' reactions to aspirin are monitored repeatedly over time. (p. 249)

Nothing said so far is meant to belittle causal inference. Whether or not we view it as involving an ineliminably *counterfactual* element, we can certainly draw safe causal conclusions when the relevant assumptions are fulfilled. Actually, both the advocates of the counterfactual approach (e.g., Holland 1986; Cox & Wermuth 2001) and their opponents (e.g., Dawid 2000) agree that we can get valuable information about the so-called *average causal effect*. This is the average causal effect on the whole population, that is, the difference between the expected value of responses to treatment t and the expected value of responses to treatment c. Indeed, randomised controlled experiments are important precisely because they let us know about average causal effects.[11] However, to get from the average causal effect in a population to the *individual causal effect* on a specific unit u, we need the further assumption of 'constant effect' (Holland 1986, 948) or 'unit-treatment additivity' (Cox 1986, 963). According to this, the effect of treatment t on each and every unit u is the same.[12] Whether this holds or not is a largely empirical matter.[13]

7.5   Using a Black Box versus Looking into it

Given the difficulties of the interventionist semantics for counterfactuals, it seems prudent to assume the simpler and more comprehensive Lewis-style semantics. In any event, the differences between the various counterfactual accounts of difference-making are less important than the common difference from the mechanistic account. We have already stressed towards the end of Section 6.4 that there is a fundamental asymmetry between dependence and mechanistic approaches: the counterfactual approach (a fortiori

---

[11] For some interesting (but manageable) complication, see Kluve (2004, 86–7).

[12] In fact, the constant effect assumption is a consequence of unit homogeneity (cf. Holland 1986, 949). For some criticism of this assumption, see Cox and Wermuth (2001, 68).

[13] The counterfactual approach to causal inference has been severely criticised by Philip Dawid (2000) and has been vigorously defended by others (see the discussion that follows Dawid's article). Dawid has a number of important complaints. But the thrust of his critique is that the counterfactual approach relies on untestable metaphysical assumptions and, in particular, on a hopeless attempt to calculate the value of an unobservable quantity. Dawid's reaction, though invariably interesting, may be too positivistic.

the *dependence* approach) is more *basic* than the mechanistic (a fortiori the *productive*) one in that a proper account of mechanisms depends on counterfactuals, while counterfactuals need not be supported (or depend on) mechanisms.

The argument is worth repeating because of its centrality. The mechanistic approaches fail to account for the interactions among the parts of the mechanism unless they assume difference-making relations under actual and counterfactual interventions. MDC are not quite clear on what the interaction within the mechanism consists in. The mechanistic approach to causation fills in the 'chain' that connects the cause and the effect with intermediate loops. But there is still no account of how the loops interact. So, the interaction between parts A and B of mechanism M is accounted for by positing an intermediate part C. But then, there is need to account for the interactions between A and C, and C and B. Then we posit other intermediate parts D and E and so on. It might well be the case that the most general and informative thing that can be said about these interactions is that there are relations of counterfactual dependence among the parts of the mechanism.

Imagine a perfectly randomised experiment in which t (for treatment) produces higher response than c (for control). Has a causal connection been established? If we treat the randomised experiment as a black box, then insofar as it is a *good* experiment, we have established a causal connection. But what is inside the *black box*? Some might think that without a specification of the mechanism by which the higher response t was effected, the causal connection has *not* been established.[14]

This is a delicate issue. As noted in the end of the last section, establishing the causal status of each part of a mechanism would require finding out (or estimating) its causal effect. And the best way to do this is by non-mechanistic means, and in particular by means of counterfactual dependence. So, there seems to be a genuine asymmetry here. The causal effect can be found out, at least in favourable circumstances, *without* understanding the causal mechanisms, if any, involved; but the causal mechanisms, even if they are present, cannot be understood without the notion of the causal effect, that is, without some notion of (counterfactual) dependence.

---

[14] Notably, this is the view of D. R. Cox (1992, 297). He claims that this was also R. A. Fisher's view. When asked, at a conference, for his view on the step from association to causation, Fisher is reported to have responded: make your theories elaborate (cf. Cox 1992, 292). It's also the thrust of the Russo-Williamson Thesis discussed in Section 4.3.2.

But there are at least three things that show how mechanistic considerations can help the counterfactual approach to causal inference. First, mechanistic considerations can help testing the stability assumptions (unit homogeneity, temporal stability) that are necessary for the counterfactual inference. We take this to be fairly obvious, so we won't elaborate on it further.

Second, mechanistic considerations can help deal with the endogeneity problem. Briefly put, the problem of endogeneity is this. It might happen that the values taken by the so-called explanatory (or causal) variable are consequences, rather than causes, of the values of the dependent variable. In a perfectly controlled experiment this cannot happen because the variables that are manipulated are the explanatory variables. But in cases where the research is qualitative or where an experiment is not possible at all, the counterfactual approach might well fail to solve the endogeneity problem. Consider one of the classic problems of the early twentieth-century social science: Max Weber's claim that a certain type of economic behaviour – the capitalist spirit – was induced by the Protestant ethic. Many social scientists have argued that this claim falls foul of the endogeneity problem. Opponents of Weber's thesis claimed that the order of dependence goes in the other direction: Europeans who already have had an interest in breaking free of the pre-capitalist mode of production might have broken free of the Catholic Church precisely for that purpose. That is, it was the economic interests of certain groups that caused the Protestant ethic and not conversely. In cases such as this, a controlled experiment is out of the question. Besides, the assessment of intuitively relevant counterfactuals will be, to say the least, precarious. But an understanding of the mechanisms at play can well help resolve the endogeneity problem. These mechanisms, we presume, include a more detailed description of the explosion of the capitalist economic activity in the sixteenth century and of the economic behaviour of certain groups, for example, in Venice and Florence or in England and Holland, which predate the emergence of Protestantism.

The third way in which mechanistic considerations can help the counterfactual approach concerns the possible confounders. In a perfectly randomised trial, the problem of confounding variables does not arise. The experimental method itself makes it very unlikely that the explanatory variable is correlated with possible confounders. But in qualitative research, or even when matching techniques are used, it is possible that the explanatory variable is correlated with a confounding variable. Take, for instance, the dependent variable to be participation in demonstrations and the explanatory variable to be the age of the participants. It might well be that a confounding variable (e.g., radicalness of beliefs) is correlated with the explanatory variable and has an influence on the dependent variable. In cases such as this, knowledge of mechanisms can help identify possible confounders and control for them. Conversely, knowledge of mechanisms can explain why the experimenter need not control for some variables (e.g., the colour of the eyes of those who participate in demonstrations).

Mechanisms cannot be the surrogate of a careful experiment. If we think of an experiment as a black box, then counterfactuals have a key role to play. After all, when certain assumptions hold, they can establish a causal relation. But without some knowledge of the mechanism inside the black box, we won't have *full* understanding of the causal relation. Nor can we solve, at least as effectively, some methodological problems of causal inference.

Using the black box carefully does establish a causal link. *Looking into* the box does offer extra understanding, even if it does *not*, in and of itself, establish the causal relation.

PART III

Beyond New Mechanism

CHAPTER 8

Constitution versus Causation

8.1 Preliminaries

New mechanists typically distinguish between causal and constitutive mechanisms, and between causal and constitutive mechanistic explanations.[1] Whereas causal mechanisms are taken to cause phenomena, constitutive mechanisms are taken to constitute them. Constitution, according to Craver's (2007a) widely adopted account, or 'constitutive relevance' as he calls it, is viewed as a non-causal dependency relation between the phenomenon and the components of the mechanism.

To illustrate what it is for a mechanism to constitute a phenomenon, Craver introduced his well-known diagrams (which have been called 'Craver diagrams'; see Figure 8.1). At the lower part of the diagram, we have the mechanism's component entities and activities that are taken to constitute the phenomenon, which is depicted as a dark oval above the mechanism. While Craver does not offer a theory of what constitution amounts to, he takes the phenomenon to supervene on the mechanism as a whole (i.e., the oval in the lower part that includes all the components of the mechanism). Craver describes the components of the mechanism as '$X_1\varphi$-ing', '$X_2\varphi$-ing' and so on, where the term '$X \varphi$-ing' refers to an entity X engaging in an activity φ. He describes the phenomenon as $S\psi$-ing, where the S is an entity that engages in the activity ψ and where $S\psi$-ing is the phenomenon that is taken to be constituted by the mechanism. To illustrate, when a neuron generates an action potential, S is the neuron, $S\psi$-ing the neuron generating an action potential, and the Xs that φ are the various components of the mechanism that is responsible for the generation of the action potential (in what follows we will adopt Craver's terminology to refer to the components of the mechanism and to the phenomenon).

---

[1] This distinction goes back to Salmon (1984).

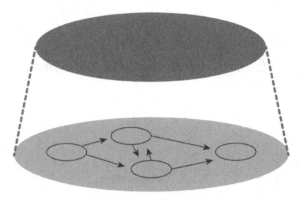

Figure 8.1   A Craver diagram (based on figure 1.1 from Craver 2007a, 7).

Importantly, not all parts of the neuron are components of the mechanism for the action potential. In general, in the case of an S that ψs, the S will typically have several parts that are not components of the mechanism responsible for S's ψ-ing. So, not all parts of S are also components of the mechanism responsible for S's ψ-ing. The components are said to be the parts that are 'constitutively relevant' to S's ψ-ing. As Craver puts it, 'the very idea of a mechanism presupposes the idea of constitutive relevance ... This difference between mechanisms and machines turns at least in part on the fact that all of the constituents of a mechanism are relevant to what it does (they are components), while only some of the parts of machines are relevant to what they do (namely, those that are components in one or more of its mechanisms)' (2007b, 6).

8.2   Craver on Constitutive Relevance

Craver gives two conditions for constitutive relevance. According to the mutual manipulability condition (1) there has to be some change to X's φ-ing that changes S's ψ-ing, and (2) there has to be some change to S's ψ-ing that changes X's φ-ing. The parthood condition says that the Xs are parts of S.[2] Craver uses Woodward's notion of ideal intervention to give an

---

[2] For Craver, then, constitutive relevance does not amount to a supervenience or identity claim. It is a relationship between the phenomenon (S's ψ-ing) and one of the components of the mechanism responsible for it. S's ψ-ing does not supervene on X's φ-ing; rather, it supervenes on the organised activities of all of the components in the mechanism. Note also that Craver says that the mutual manipulability account provides a sufficient condition for something to be a component of a mechanism; but he also says that the failure of mutual manipulability is a sufficient condition for

account of mutual manipulability. As we saw in Chapter 7, an ideal intervention I on X with respect to Y changes the value of Y only via the change in X. Craver defines an analogous notion of ideal intervention for constitutive mechanisms. The main idea is that 'an *ideal* intervention I on φ with respect to ψ is a change in the value of φ that changes ψ, if at all, *only via* the change in φ', and similarly for an ideal intervention on ψ with respect to φ (2007a, 154). So, for φ to be constitutively relevant to ψ, '[t]here should be some ideal intervention on φ under which ψ changes, and there should be some ideal intervention on ψ under which φ changes' (p. 154).

The adequacy of Craver's account is a contested issue in the mechanistic literature. A main problem for this account is whether it is indeed possible to apply Woodward's notion of ideal intervention to the case of constitutive relevance. It is not possible to intervene on one of the components of S's ψ-ing without also intervening on S's ψ-ing, and vice versa. As Michael Baumgartner and Lorenzo Casini (2017) note, an intervention I on ψ with respect to φ must be viewed as the common cause of the changes in both ψ and φ, and so is not an ideal intervention on ψ with respect to φ, in contrast to what Craver assumes. Baumgartner and Casini (2017, 220) conclude that 'the types of interventions required by [mutual manipulability] cannot possibly exist for any mechanistic system. [Mutual manipulability] is hence unsatisfiable, which means that constitutive relations as defined by [mutual manipulability] are inexistent, which again entails that friends of mechanistic explanations who rely on [mutual manipulability] chase a chimera.'[3]

If Craver's account of constitutive relevance is problematic, what options are open for the mechanist? Craver holds both that constitutive relevance is needed in order to understand (constitutive) mechanisms and that this can be done in terms of interventionism. A first option, then, is to retain both claims, but to try to find an alternative way to reconcile

---

something to fail to be a component (2007a, 159). As Baumgartner and Casini note (2017), this amounts to a necessary and sufficient condition for something to be a component of a mechanism.

[3] Several other criticisms have been made to Craver's account. Leuridan, for example, has argued that mutual manipulability fails to provide a sufficient account for constitutive relevance, since 'plenty of cases of bidirectional causation would unintentionally fit the mutual manipulability account of constitutive relevance' (2012, 400). Franklin-Hall (2016) has argued that although Craver's account can show whether a part of a whole is relevant to some behaviour of the whole, it does not solve what she calls the 'carving problem', that is, how to provide appropriate decompositions of mechanisms. Another concern is that Craver's account seems to offer a way to decide whether a part is a component of a mechanism, but does not tell us what constitutive relevance really is (Couch 2011). This is also Glennan's main complaint (2017, 44).

interventionism with mechanistic constitution.[4] A second option is to retain the first claim, but deny that interventionism can be used to give an account of constitutive relevance and attempt to find an alternative account.[5] There exists a third, more radical, option, that is, to deny that constitutive relevance, qua a non-causal dependency relation, is in fact needed in order to understand what a mechanism is. This (unpopular) option leads to a deflationary view of 'constitutive' explanation, as being a certain kind of causal explanation, and is the view that will be defended in this chapter.

8.3 Are There Constitutive Mechanisms?

Craver first introduced the 'Craver diagram' representation of a (constitutive) mechanism in *Explaining the Brain* (2007a). He begins with the example of the mechanism of neurotransmitter release by a neuron. Consider Craver's description of this mechanism:

> The mechanism begins, we can say, when an action potential depolarizes the axon terminal and so opens voltage-sensitive calcium (Ca^{2+}) channels in the neuronal membrane. Intracellular Ca^{2+} concentrations rise, causing more Ca^{2+} to bind to Ca^{2+}/Calmodulin dependent kinase. The latter phosphorylates synapsin, which frees the transmitter containing vesicle from the cytoskeleton. At this point, Rab_3A and Rab_3C target the freed vesicle to release sites in the membrane. Then v-SNARES (such as VAMP), which are incorporated into the vesicle membrane, bind to t-SNARES (such as syntaxin and SNAP-25), which are incorporated into the axon terminal membrane, thereby bringing the vesicle and the membrane next to one another. Finally, local influx of Ca^{2+} at the active zone in the terminal leads this SNARE complex, either acting alone or in concert with other proteins, to open a fusion pore that spans the membrane to the synaptic cleft. (pp. 4–5)

Craver goes on to offer a general characterisation of a mechanism by saying that the mechanism of neurotransmitter release 'is a set of entities and

---

[4] See, for example, the accounts given in Romero (2015) and Harinen (2018) – we will discuss Harinen's account in Section 8.5. One possibility is to adopt a modified form of interventionism on which it is not required, when intervening on some variable, to hold other variables fixed, if these are related by, for example, mereological or supervenience relations to the variable on which the intervention is applied (see Woodward 2015). Adopting such a framework makes interventions on variables related by constitutive relations possible, but it seems to lead to a problem of empirical underdetermination when we try to infer constitutive relations (see Baumgartner & Casini 2017).

[5] See, for example, the 'regularity' accounts of constitutive relevance offered in Harbecke (2010) and Couch (2011); see also Gillett (2013) for an account based on the dimensioned view of realisation.

activities organized such that they exhibit the phenomenon to be explained' (p. 5). The word 'exhibit' here is meant to capture the specific relation between the organised entities and activities, on the one hand, and the phenomenon, on the other; the organised entities and activities, that is, the mechanism, do not *produce* the phenomenon, since they are not its antecedent causes, but they *exhibit* it. Craver then offers his well-known diagram for the abstract representation of a mechanism and notes that this account leads to a type of explanation different from 'etiological causal explanation' which says 'how a phenomenon is produced by its causes', namely, to 'constitutive (or componential) causal-mechanical explanation', which is 'the explanation of a phenomenon ... by the organization of component entities and activities' (p. 8).

According to us, the mechanism of neurotransmitter release is an example of a mechanism as a causal pathway; similar to the case of the signalling pathways of apoptosis, we have a sequence of causal steps that causally link the depolarisation of the axon terminal to the opening of the fusion pore and the release of neurotransmitters.[6] Such a case, then, as well as the cases of biological pathways we examined in Chapter 3, cytological cases like apoptosis or mitosis, pathological mechanisms, cases like the mechanism of DNA replication and protein synthesis often discussed by mechanists and the several other examples we have seen in the book, are best viewed, we think, as causal sequences that produce an effect, and not in terms of a 'Craver diagram'.

Given the prevalence of mechanisms as causal pathways it is perhaps strange that constitutive mechanisms have become central among new mechanists. It is important to note that mechanists recognise that some mechanisms are best viewed as causal sequences. Although this is perhaps not very explicit in Craver (2007a), new mechanists commonly distinguish between two types of mechanisms. For example, Craver and Darden (2013, 65–6) distinguish between 'productive mechanisms' and 'underlying mechanisms', where in each case there is a different kind of relationship between the mechanism and the phenomenon. In the case of productive mechanisms 'one typically starts with some understanding of the end product and seeks the components that are assembled and the processes by which they are assembled and the activities that transform them on the way to the final stage'. In the case of underlying mechanisms, in contrast, 'one typically breaks a system as a whole into component parts

---

[6] Craver seems to agree, as he later (2007a, 22) notes that this is an etiological explanation, where the explanans 'is the mechanism linking the influx of Ca^{2+} into the axon terminal'.

that one takes to be working components in a mechanism, and one shows how they are organized together, spatially, temporally, and actively such that they give rise to the phenomenon as a whole'. Productive mechanisms are similar to mechanisms as causal pathways (although of course we would not characterise causal pathways in terms of productivity or in terms of the distinction between entities and activities, as we think that such talk unjustifiably metaphysically inflates the notion of mechanism).

Similarly, Craver and Tabery (2015) note that some mechanisms can be conceived 'as a causal sequence terminating in some end product: as when a virus produces symptoms via a disease mechanism or an enzyme phosphorylates a substrate'. But not all mechanisms are to be viewed in such terms. They claim that for 'many physiological mechanisms ... it is more appropriate to say that the mechanism *underlies* the phenomenon'. Examples are the mechanism of the action potential and the mechanism of working memory. The phenomenon here is not the production of something, but 'a capacity or behavior of the mechanism as a whole'.

New mechanists, then, can accept that at least some mechanisms in biology are better viewed as causal pathways. The question, then, is not whether mechanisms as causal pathways exist (or whether mechanism talk in life sciences is to be understood in terms of causal pathways), but whether we have to accept a second type of mechanism, that is, constitutive mechanisms. It is not our claim that the notion of a constitutive mechanism is conceptually problematic (although we think that a clear account of what constitutive relevance amounts to is still lacking). For us, the main question is: Is the notion of a constitutive mechanism important in illuminating the notion of mechanism as a concept-in-use central in practice? Given the prima facie adequacy of Causal Mechanism to capture this concept, what would be a convincing argument to accept that at least some instances of the notion are to be understood in terms of constitutive mechanisms? Ideally, we need a more general reason than simply to point out that a limited number of cases are better viewed in such terms.

Is, then, a 'constitutive' sense of mechanism present in scientific practice? A point to be noted here is that this is not an easy question to answer if we don't know what constitution or constitutive relevance is. For example, Craver notes that mechanistic constitution is not material constitution, which is a notion about which we may have various intuitions. And as we have seen, Craver's account of it is not without problems. But then we face the following difficulty: If we do not have a clear account of what a constitutive mechanism is, how exactly can we look for it within scientific practice? A solution could be to examine what scientists themselves say about mechanistic constitution or to look at some paradigmatic cases of mechanistic constitution that are regarded as such by scientists in order to try to analyse this notion. One might think that the situation here is similar to what we do when we want to analyse the notion of causation as it functions within scientific practice: for instance, we look at the structure of experiments performed to reveal causal relations. The difference-making account of causation, for example, is grounded in this way in scientific practice, as we saw in Chapter 4 with the case of scurvy. But there is an asymmetry here: in contrast to the case of causation or mechanism in the sense of CM, it is not easy to argue that there is a notion of mechanistic constitution present in scientific practice.[7]

Be that as it may, we can use Craver's account of what a constitutive mechanism is, even if it is not entirely clear what mechanistic constitution amounts to, to try to see whether something like this can be found in scientific practice. We can identify two strategies to provide such a general reason in order to argue that constitutive mechanisms are important. Both strategies attempt to show that if our account of mechanism as a concept-in-use is confined to viewing a mechanism as a causal pathway, we fail to capture a central aspect of biological practice; that is, our account fails to be descriptively adequate.

The first strategy is simply to say that when we look at scientific practice, we find many instances of constitutive mechanistic explanations; to understand how exactly these explanations function we need to view a mechanism in 'constitutive' terms, that is, as constituting (or underlying) the phenomenon to be explained. The second strategy is to say that there are certain kinds of experiments in biology that can be understood only by adopting a 'constitutive' account of mechanisms. If these strategies succeed, the account offered by CM leads to an impoverished notion of mechanism as a concept-in-use since it fails to capture some salient features of scientific practice.

---

[7] This emphasis on scientific practice when offering an account of mechanisms is a central feature of Baetu's (2019) analysis. In relation to Craver's constitutive relevance account, he says: 'Instead of first committing to a noncausal interpretation and then attempting the impossible task of demonstrating that evidence for causation can somehow be used to demonstrate something else than causation, one can simply let go of ready-made metaphysical intuitions borrowed from physics and the philosophy of mind and consider the relationship between mechanisms and phenomena from the strictly experimental standpoint' (p. 33). He thinks that this 'strictly experimental standpoint' leads to what he calls the 'causal mediation account' concerning the relationship between mechanisms and phenomena. Baetu's analysis, with its emphasis on methodology, shares many similarities with CM.

An example of the first strategy can be found in Marie Kaiser and Beate Krickel's (2017) analysis of constitutive mechanisms. According to them,

> one can etiologically explain protein synthesis by describing how a certain sequence of causes leads to the synthesis of a protein, or one can constitutively explain protein synthesis by referring to the components of a cell and describing how they act and interact such that the cell synthesizes proteins. On a closer inspection, however, it turns out that what we are explaining is not the same phenomenon, but two different phenomena: the etiological MEx [mechanistic explanation] explains the end-result (there being a protein) and the constitutive MEx explains the process of protein synthesis (we want to know what happens at every step of protein synthesis). (p. 8)

The claim here is that a constitutive mechanistic explanation explains something that an etiological mechanistic explanation fails to explain. If this is correct, then we have a general reason why constitutive mechanisms are central: they explain an aspect of the phenomenon that cannot be explained otherwise. On this view, constitutive mechanisms seem to be as important and ubiquitous as mechanisms qua causal sequences: when we have a causal sequence mechanism, we have a constitutive one, and vice versa.

But this does not sound very plausible. Consider again protein synthesis; what more is there to be explained, if we know everything that is to be known about the causal sequence that leads to the production of the protein? And what more is there to be explained if we know all the causal steps in a metabolic or signal transduction pathway? A mechanistic explanation in our sense, that is, a description of the causal pathway of protein synthesis, is sufficient to explain 'the process of protein synthesis' or 'what happens at every step' of the mechanism.

The second strategy is exemplified by Craver's analysis of what he calls 'interlevel' experiments. As he puts it, '[t]he norms of constitutive relevance are implicit in the experimental strategies that neuroscientists use to test claims about componency and in the rules by which neuroscientists evaluate applications of those strategies' (2007a, 144). Interlevel experiments for Craver can be bottom-up or top-down. In bottom-up experiments one intervenes to change the components of the mechanism and detects how the phenomenon (Sψ-ing) changes. In top-down experiments the intervention is applied to the phenomenon and one then detects how the behaviour of the components of the mechanism change. Craver uses the bottom-up/top-down distinction and the distinction between an inhibitory and an excitatory intervention to classify the most common kinds of experiments in neuroscience, i.e., interference, stimulation and activation experiments (pp. 144–52). Interference experiments are bottom-up and inhibitory (e.g., lesion experiments), stimulation experiments bottom-up and excitatory (e.g., stimulating the brain to produce movements in specific muscles) and activation experiments top-down and excitatory (e.g., engaging the subject in some task so that a cognitive system gets activated and monitoring what happens in the brain as in PET and fMRI studies). Craver reviews various inferential challenges that these experiments face; the combined use of these experiments, that is, performing both a bottom-up and a top-down experiment, is a way to overcome these challenges. For example, in an interference experiment a lesion may lead to a change in the phenomenon, but this does not necessarily mean that a particular brain region is a component in the mechanism responsible for the phenomenon; an alternative hypothesis is that the disrupted brain region leads to a change in some other part of the brain, and this other part is a component in the mechanism.

These sort of challenges and the ways they are overcome are used by Craver 'as data points in building a descriptively and normatively adequate account of constitutive relevance' (p. 147) and provide a main motivation for his mutual manipulability account. His strategy, then, is to use interlevel experiments to offer an adequate account of what it is for X to be a component of a mechanism. Craver seems to take it for granted that interlevel experiments are to be viewed in terms of 'Craver diagrams' (p. 146); he thus applies a 'constitutive' framework to analyse the experiments and asks what they suggest about the norms of constitutive relevance. But one may use these experiments to justify the 'constitutive' framework itself and the salience of constitutive mechanisms in scientific practice. The argument here, then, is that only 'constitutive' mechanisms can make sense of interlevel experiments; the view of mechanisms as causal pathways cannot capture this important feature of scientific practice. The task for us, then, is to show that Causal Mechanism can lead to an alternative account of interlevel experiments. This alternative account will involve a different view of levels than Craver's; we will introduce the main idea of this alternative account in this chapter, but we will develop it more fully in Chapter 9.

8.4 Against Constitutive Mechanisms

We have seen that new mechanists think that in certain cases it is more appropriate to view mechanisms in constitutive terms. But when exactly is it more appropriate to do so? In a constitutive mechanism the organised entities and activities explain the behaviour of some entity of which they are parts, that is, what Craver refers to as 'Sψ-ing' in his diagram. Our question thus becomes: When is it appropriate to talk about an S that ψs?

A natural suggestion is that we apply such language when there are natural boundaries around a mechanism, that is, when the mechanism is part of a biological object or structure. For example, the metabolic pathway of glycolysis occurs inside the cell, so we can take the cell as the S in this case.[8] A problem here is that many mechanisms are not confined within natural boundaries. For example, in signal transduction pathways the signal comes from outside the cell (as in the case of the extrinsic pathway of apoptosis), and so, although a component of the pathway, it is not part of the cell. Kaiser and Krickel (2017, 29) mention the example of the mechanism for muscle contraction which also includes an external signal as one of its components, that is, 'neurotransmitters that bind to receptor molecules outside of the muscle fibre'; another example is from David Kaplan (2012, 552), the gecko adhesion mechanism, which is 'spatially distributed to include external components spanning the boundary between the gecko and its environment'. In pathological mechanisms, too, the causal agent of a disease can be an environmental factor, for example, radiation, extreme temperature, toxic substances or viruses.

Craver is aware of this point; as he says, 'mechanisms frequently transgress compartmental boundaries' (2007a, 141). His main example, the mechanism of the action potential, is a case in point. This mechanism 'relies crucially on the fact that some components of the mechanism are inside the membrane and some are outside. The membrane allows the intracellular and extracellular concentrations of ions to be different, allows a diffusion gradient to be set up, and allows for a separation of charge' (p. 141). Also, cognitive mechanisms 'draw upon resources outside of the brain and outside of the body to such an extent that it may not be fruitful to see the skin, or the surface of the CNS, as a useful boundary' (p. 141).

---

[8] Of course, if a mechanism such as glycolysis occurs inside a cell, there will be several other biological structures of which it is a part, for example, the tissue in which the cell is a part, the organ in which the tissue is a part, and finally the organism. However, it is more natural in this case to regard the cell as the S, the behaviour of which the mechanism explains. In some cases, however, it may not be clear which biological structure should be regarded as the S; for example, what about a mechanism that occurs inside a cell organelle? DNA replication in eukaryotes occurs in the nucleus; is the nucleus or the cell the S in this case? Moreover, it may seem strange to attribute some behaviour to the cell itself in virtue of the fact that DNA is replicated inside its nucleus, or some protein is synthesised, or glycolysis occurs. It is, we think, more appropriate to view all these processes as pathways that occur inside a cell, without taking the cell as an acting entity and viewing the pathways in terms of a constitutive mechanism that involves the behaviour of the cell as a whole.

We have mentioned earlier that for Craver the components of a mechanism have to be parts of the entity S. In view of the previous examples, however, this is problematic; in all these examples, some components of the mechanism are not parts of the entity S.[9] Kaiser and Krickel (2017) suggest weakening the parthood condition, that is, to require most, but not all, of a mechanism's components to be parts of S. In this case, we will still have a constitutive mechanism, that is, an S that contains part of a mechanism that explains some behaviour of S.[10]

We draw a different lesson from the existence of components of a mechanism that are external to the entity S. Let us ask the following question: When is it appropriate to apply constitutive terms to a specific example of a mechanism? In other words, when is it appropriate to view a particular mechanism in terms of a Craver diagram? We think that the existence of components that are external to a biological structure that is a candidate for being the entity S undermines treating this biological structure as the entity that ψs; hence, we should either look for some other entity or treat the mechanism not as a constitutive one but as a causal sequence. It is not clear, however, which other biological structure can be the entity S; even if the whole organism is taken as the S (which is implausible for a mechanism such as a signal transduction pathway), we have seen that there are mechanisms the components of which are external to the organism. If, alternatively, the S is taken as the mechanism as a whole, as Craver seems to suggest, this leads to problems (as we mention in note 9).

The problem of external components, then, is not a reason to reject or weaken the parthood condition of constitutive relevance but a reason to view the mechanism in terms of a causal sequence and not in 'constitutive' terms. We take typical and paradigmatic biological mechanisms to be causal pathways, where these pathways are not necessarily confined within specific biological objects or structures, but can transgress natural boundaries. In many cases of pathways, that is, there will not exist an appropriate

---

[9] Although we take S to be an entity such as a neuron that contains the components of a mechanism that explains some behaviour of S, Craver also says that S refers to 'the mechanism as a whole' (2007a, 7). We agree with Kaiser and Krickel (2017, 26) that this latter claim is problematic; as they say, '[i]t is not the mechanism of muscle contraction that is contracting. Nor is it the action potential mechanism that fires, or the spatial memory mechanism that navigates through the Morris Water Maze. Rather, the relevant objects are individuals that in most cases are larger objects or systems that contain one or often more mechanisms.'

[10] Note that the option to reject entirely the parthood condition of constitutive relevance leads to problems, as cases of causal feedback mechanisms would count as examples of constitutive relevance (cf. Leuridan 2012).

biological object S, where every component of the pathway is part of S. Such mechanisms are better viewed in etiological, rather than constitutive, terms.[11]

Let us now consider a mechanism, for example, protein synthesis, all the components of which are parts of a biological object, that is, a cell. Would we be justified to regard the cell as the S that ψs? A problem here is that the mechanism of protein synthesis, as well as several other mechanisms, can exist outside the cell. In fact, this is how the molecular details of these mechanisms were discovered. This process involves what are known as cell-free systems; to develop a cell-free system cells must be disrupted and fractionated. The mechanism of protein synthesis was first studied by using a cell homogenate that translated RNA molecules, producing proteins. As Alberts et al. (2014, 451) explain, '[f]ractionation of this homogenate, step by step, produced in turn the ribosomes, tRNAs, and various enzymes that together constitute the protein-synthetic machinery'. Individual purified components 'could be added or withheld separately to define its exact role in the overall process'. As the authors stress,

> [a] major goal for cell biologists is the reconstitution of every biological process in a purified cell-free system. Only in this way can we define all of the components needed for the process and control their concentrations, which is required to work out their precise mechanism of action ... [A] great deal of what we know today about the molecular biology of the cell has been discovered by studies in such cell-free systems. They have been used, for example, to decipher the molecular details of DNA replication and DNA transcription, RNA splicing, protein translation, muscle contraction, and particle transport along microtubules. (p. 451)

The fact that a mechanism such as protein synthesis can operate outside a cell means that the components of the mechanism (the Xφ-ings) can exist, and the mechanism can function, without the entity that they constitute (the Sψ-ing). But is this possible? The organised entities and activities that are constitutively relevant to the phenomenon P (the mechanism for P) are taken to constitute the phenomenon (cf. Kaiser & Krickel 2017, 9). But it seems strange to have a mechanism that, if present in a cell, constitutes phenomenon P, but if outside the cell, although it operates

---

[11] But is there a fact of the matter about where exactly the starting point of a mechanism is? If it is more or less arbitrary what we take as the starting point of a causal sequence, why not just focus on the part of a causal sequence that is part of an entity S? That way we may avoid the problem of external components. Our answer is that what is arbitrary is to insist that all the components of a mechanism have to be parts of the entity S; as we have seen, paradigmatic mechanisms transgress natural boundaries.

normally, does not constitute anything or perhaps constitutes something different.[12] Moreover, if we take S's ψ-ing to supervene on the mechanism for S's ψ-ing in the sense that 'there can be no difference in S's ψ-ing without a difference in the mechanism for S's ψ-ing' (Craver 2007a, 153, n. 33), then in the case of cell-free protein synthesis we have a difference in S's ψ-ing (trivially, since we have no cell that ψs – and we cannot have just the ψ-ing without the entity that engages in it) but no difference in the underlying mechanism. The conclusion, then, is that the entity S in such a case cannot be the cell; it is best to view the mechanism as a causal pathway, without also having an S that ψs.[13]

We take the arguments in this section to show that it is not appropriate or useful to view typical and paradigmatic biological mechanisms in constitutive terms. Paradigmatic mechanisms such as the mechanism of protein synthesis and metabolic pathways are not constitutive mechanisms to be represented by Craver diagrams, and many of them transgress natural boundaries. Biological mechanisms are better viewed, then, as pathways and analysed in terms of causation (and not constitution).

8.5 Causal Mechanism and Constitutive Relevance

We will close this chapter by arguing that some of the elements of Craver's analysis still hold if we adopt the view that mechanisms are causal pathways. Consider again the distinction between etiological causal-mechanical explanation and constitutive or componential causal-mechanical explanation. This distinction can be retained, even if we reject Craver's account of constitutive mechanisms. In the case of an etiological explanation we focus on the antecedent causes of a phenomenon. In componential explanation we start with a previously established causal relationship between X and

---

[12] In the case of material constitution, if A materially constitutes B, it is possible for A to exist without constituting B; but as Craver stresses, mechanistic constitution is not material constitution, so we have no reason to expect that an analogous claim holds in the case of mechanistic constitution.

[13] Another option is to take as the entity S not the cell, but the mereological sum of all the components of the mechanism. For example, in the case of the protein synthesis the S could be the mereological sum of the DNA and RNA molecules, the amino acids, various enzymes and so on. But this seems implausible: a mereological sum of all the components of a biological pathway is not a 'natural' biological object, and it seems very strange to attribute a behaviour to that object,' which the mechanism is supposed to explain. In general, it is not plausible to view a mechanism as a whole as an object or as an acting entity. Mechanisms, to use a distinction from metaphysics that Kaiser and Krickel apply in their analysis, are better viewed as occurrents, that is, things that occur like football matches and weddings, and not as continuants, that is, things like stones and tables. But if mechanisms are occurrents, it again seems strange to think that they constitute the ψ-ing of an entity S, where S is presumably a continuant, for example, a cell.

Y and focus on the intermediate steps, that is, the components, of the causal pathway that leads from X to Y.

As we have noted, this is how a mechanism is viewed in statistics; that is, it concerns a causal structure that links a cause with an effect variable. In our example from Chapter 4, vitamin C is the mediating variable in the causal relationship between citrus fruits and scurvy. Pathological mechanisms, in general, link a causal agent such as a virus to the development of a disease. Similarly, the signalling pathways of apoptosis link an initial trigger such as a toxic agent to the cytological process of apoptosis described by Kerr et al. (1972). Identifying the components of a causal pathway that links an initial cause to some effect can then be usefully distinguished from etiological causal explanation that identifies antecedent causes, although in both cases what we are identifying are causal relationships between variables or components of a pathway. Mechanism talk is more appropriate if what we are interested in is finding the mediator or, more generally, 'extending' a causal pathway by identifying intermediate causal steps. This is because by a 'mechanism' biologists typically mean the way a cause produces the effect. In such contexts, causal explanation can be viewed as 'componential' and not a mere matter of identifying some antecedent cause or establishing a causal relationship between two variables.

However, several other claims that Craver makes will have to be rejected. In particular, we do not accept a non-causal relation of constitutive relevance; components of pathways are only *causally* relevant to the outcome of the mechanism. We can perhaps say things like 'the components of the signalling pathway of apoptosis constitute the pathway', but this should be viewed in a deflationary way and not as referring to some synchronic and symmetric relationship between relata that are not wholly distinct (as Craver characterises the relationship of constitutive relevance). In fact, we do not think that it is very useful (and it makes much sense, at least in the case of the typical and paradigmatic cases of biological mechanisms we have been considering) to talk about 'the behaviour of a mechanism as a whole'. This kind of talk can lead to a metaphysical inflation of the notion of mechanism, where we think of the mechanism itself as a kind of entity that engages in some activity (an S that ψs) and then ask about the nature of the relation between the components of the mechanism and S's ψ-ing. We have argued in this chapter that this general picture is not appropriate when we look at typical biological mechanisms.

Attempts to view the relation of constitutive relevance in causal terms lead to accounts that are not very far from the picture we want to defend (see especially Menzies 2012 and Harinen 2018). The main idea here is that S's ψ-ing can be viewed as an input-output relationship where the mechanism is causally between the input and output.[14] Totte Harinen, in particular, represents the phenomenon that the mechanism exhibits 'as a causal relation holding between two variables, ψ_{in} and ψ_{out}, corresponding to the input and output conditions characteristic of the relevant regularity' (2018, 14). On this view, in assessing mutual manipulability we have to take three variables into account: the higher-level variables ψ_{in} and ψ_{out}, which are the input and output variables of the mechanism, and a lower-level variable φ_i that is causally between ψ_{in} and ψ_{out}. In a top-down intervention we intervene on the input variable ψ_{in} and observe whether the lower-level variable φ_i changes. In bottom-up experiments we intervene on the lower-level variable φ_i and observe whether the output variable ψ_{out} changes. In claiming that constitutive relevance can be viewed as causal relevance, Harinen accepts interlevel causation, which is denied by several philosophers who think that there can be no interlevel causal relations between a mechanism (or Sψ-ing) and its components (cf. Craver & Bechtel 2007; Romero 2015; Baumgartner & Gebharter 2016).[15]

We agree with Harinen that mechanisms can include interlevel causation, but we do not view 'levels' in terms of Craver's levels of mechanisms; levels for us are levels of composition (or organisation), where levels and mechanisms are distinct notions. In Chapter 9 we will develop this view in some detail and claim that it can offer an alternative account of interlevel experiments. Our view, then, is that causal pathways can contain components from various levels of organisation, and representations of mechanisms can contain both higher-level and lower-level variables. So, we can have pathways like those suggested by Harinen ($\psi_{in} \rightarrow \varphi_i \rightarrow \psi_{out}$),

---

[14] Menzies (2012) uses a structural equations framework to provide an analysis of the causal structure of a mechanism 'in terms of the composition of functional dependences, or, in material mode, in terms of the programmed exercise of modular capacities' (p. 804). Although he rejects Craver's account of constitutive relevance, he thinks of his own account as retaining a notion of constitutive relevance distinct from causal relevance. For him 'any variable that lies on a pathway between the input variable and the output variable of the capacity to be explained counts as part of the mechanism underlying the capacity' (p. 801). So, intervening variables on a pathway 'are constitutively relevant to the mechanism by virtue of being parts of the pathway that is the mechanism'. But these variables 'are also causally relevant to the input and output variables'. Menzies thus takes the two relevance relations to be different. Whereas talk of systems having capacities that are explained by the mechanism may be appropriate for psychology and cognitive science, we are sceptical whether this is a useful interpretation of the biological examples we have been considering in this book.

[15] Harinen's account uses the modified interventionist framework developed in Woodward (2015) and so does not have the problem that Craver's original account had in relation to the feasibility of ideal interventions.

where the input and the output variables are at a higher level of organisation, whereas the other components are at lower levels; in the case of scurvy, for example, the mediator (vitamin C) is at a lower level of organisation than the input and output variables (a diet of citrus fruits and scurvy). But it is also possible to have other kinds of pathways, with various combinations of levels (in the next chapter we will review various such examples).

So-called constitutive explanation is for us a version of causal explanation, and there is no need to posit non-causal relations such as constitution, constitutive relevance or supervenience in order to understand what a mechanism is in biology. To do so is, as in the case of general characterisations of mechanisms that refer to entities and activities, to inflate the notion of mechanism as concept-in-use. Importantly, as we will see, such a view of mechanisms leads to a different understanding of the notion of levels and the relationship between levels and mechanisms from the dominant view among new mechanists, which is the account of levels of mechanisms developed by Craver.

CHAPTER 9

Multilevel Mechanistic Explanation

9.1 Preliminaries

There are various notions of 'levels' in science and philosophy (see Craver 2007a); we take it, however, that a main use of the notion of 'levels', especially in the life sciences, is to refer to levels of organisation or composition, where entities at one level are parts of or compose entities at a higher level (cf. Eronen & Brooks 2018). Consider, for example, the biological hierarchy: smaller molecules, for example, nucleotides, compose bigger ones, for example, DNA; DNA proteins and other molecules are parts of cells; cells compose tissues; tissues compose organs; organs are parts of organisms, which in turn are parts of ecosystems. This is a traditional view of levels in science (cf. Oppenheim & Putnam 1958). On this view, it is a more or less objective matter on what level an entity is, as level-membership is grounded in mereological or compositional relations that are taken to be objective features of the world.[1]

We can now introduce another notion of 'levels', what we might call levels of scope. Levels of scope concern the sense in which we say that physics has a wider scope than biology. Where levels of composition concern entities composing other entities, levels of scope concern the laws of nature governing entities at various levels. Thus, laws of physics have a much wider scope than laws of biology, for example (or whatever plays the role of laws in biology), since laws of physics govern entities at all levels of the compositional hierarchy, whereas laws of biology govern specifically biological entities. Laws of special sciences, then, have a narrower scope; whereas physical laws that range over all entities have the widest scope possible. In other words, physical laws range over entities at many (and perhaps all) levels of composition, but biological 'laws' range

---

[1] We take it to be an open question whether there exists a fundamental level or, alternatively, whether there is an infinite hierarchy of levels, where every entity is composed of more basic entities.

over entities at particular levels of composition. Levels of composition give us one kind of hierarchy; levels of scope give us another. That is, we can arrange scientific disciplines according to their levels of scope, that is, the generality of their laws (or of whatever plays the role of laws).[2]

We take this picture of levels of composition or organisation to be a picture that emerges from our best science. Scientists have revealed a world of entities of diverse sizes and complexity, ranging from fundamental particles, at one end of the spectrum, to superclusters of galaxies, at the other end, in terms of size, and from fundamental particles to brains and organisms, in terms of complexity. Thus, we take levels of composition and levels of scope, irrespective of the specific topology of the structure to which they give rise, to be an uncontroversial picture.

9.2   Levels of Composition in Biology

Levels of composition are, moreover, a typical way to understand talk of levels in science, and in particular in biology. In life sciences, one often talks about 'levels of organisation' or 'levels of complexity', with each level being characterised by its distinctive sets of entities and rules of operation. We have, for example, the genetic level, where one focuses on genes, how they are expressed and so on, and the anatomical level, a higher level that involves parts of organisms such as the brain and the spinal cord. Although the behaviour of the entities at some level depends on what happens at lower levels, biologists typically think that each level is characterised by its own principles not reducible to the levels below.

By way of illustration, consider the following quotation from a well-known textbook on developmental biology by Scott Gilbert:

> The properties of a system at any given level of organization cannot be totally explained by those of levels 'below' it. Thus, temperature is not a property of an atom, but a property that 'emerges' from an aggregate of atoms. Similarly, voltage potential is a property of a biological membrane

---

[2] These hierarchies need not give rise to a simple linear structure. In Oppenheim and Putnam's account, we have indeed a single hierarchy of levels of composition, with every level of composition corresponding to a scientific discipline for which it constitutes its subject matter. But the hierarchies can be part of a much more complex structure, such as the one given in Wimsatt (1976). In such more complex structures, we still have mereological and compositional relations with different entities being governed by different kinds of laws. So, even in such structures, which give a much more realistic picture than Oppenheim and Putnam's hierarchy, it is still possible to have levels of composition and levels of scope. Note also that even if we agree with new mechanists that the life sciences are structured around mechanisms rather than laws, it does not follow that there are no laws in biology; see, for example, Waters (1998) and Mitchell (2000) for accounts of biological laws.

but not of any of its components. Higher-level properties result from lower-level activities, but they must be understood in the context of the whole. (2010, 618)

Gilbert goes on to generalise this picture:

> Parts are organized into wholes, and these wholes are often components of larger wholes. Moreover, at each biological level there are appropriate rules, and one cannot necessarily 'reduce' all the properties of body tissues to atomic phenomena.... When you have an entity as complex as the cell, the fact that quarks have certain spins is irrelevant. This is not to say, however, that each level is independent of those 'below' it. To the contrary, laws at one level may be almost deterministically dependent on those at lower levels; but they may also be dependent on levels 'above'. (p. 620; emphasis added)

Gilbert (p. 620) quotes Joseph Needham (1943), who wrote:

> The deadlock [between mechanism and vitalism] is overcome when it is realized that every level of organization has its own regularities and principles, not reducible to those appropriate to lower levels of organization, nor applicable to higher levels, but at the same time in no way inscrutable or immune from scientific analysis and comprehension.

We see here, first, the notion of level developed above: a membrane is at a higher level of organisation than the molecules that compose it. Second, entities at different levels can have properties that none of its constituent parts has (e.g., temperature, voltage potential). Third, these higher-level properties are governed by a specific set of laws or rules and are taken to be in some sense irreducible to the levels below it. We will come back to this last point about the irreducibility of levels and to how it can be made more precise in Section 9.6. For now, the main point is to establish that the notion of levels of composition (or organisation) is an important notion in biology. Moreover, this fact can be used to argue that the world itself has a level-structure: levels of composition or organisation reflect how the world is organised. We take, then, the main argument in favour of the existence of levels of composition to be an a posteriori one: namely, this is how the world according to our best science is structured.

An important question here is whether this picture leads to some form of anti-reductionism or whether levels of composition are compatible with a reductionistic picture. Despite Gilbert's and Needham's apparent rejection of reductionism, we do not think that to accept that levels of organisation are real automatically leads to an acceptance of anti-reductionism or some form of ontological emergence. One can then be a realist about levels but remain agnostic about the ultimate truth of reductionism. Thus, Hilary Putnam and Paul Oppenheim wrote about ontological levels but at the same time adopted the micro-reduction of each whole at some ontological level to its constituents. More generally, the discovery that entities are typically composed of other entities does not of itself entail an account about the best way to understand the ontological nature of levels.

What then does the thesis of the reality of levels of composition amount to? What is the ontology that a realist about levels commits to, if not to some form of anti-reductionism? Our answer is that, even if the realist about levels can remain agnostic about the issue of reductionism versus anti-reductionism, the thesis is strong enough to rule out various ontological accounts. For example, vitalism – the thesis that some wholes have vital powers that do not in some sense depend on the constituents of the whole – and dualism are ruled out. In addition, the realist about levels takes it that wholes really exist and thus rejects an eliminativist stance about higher-level entities. Hence, although minimal as an ontological position, realism about levels has enough content to rule out some ontological accounts. At the same time, it leaves open the issue of exactly how properties of wholes depend on properties of their parts.[3]

9.3   Craver on Levels of Mechanisms

We take it to be a very plausible view, to say the least, that levels of composition constitute a typical sense of levels within life sciences. According to a very influential account by Craver (2007a), levels of composition in biology, and in particular in neuroscience, are to be viewed in terms of levels of mechanisms. Craver (2007a) has developed an account of multilevel explanation in neuroscience that is based on his account of what a mechanism is in biology. Below, after briefly presenting the main features of this account, we argue for an alternative conception of multilevel mechanistic explanation.

In explaining how we are to understand levels of mechanisms, Craver and Bechtel (2007, 548) write:

> In levels of mechanisms, an item X is at a lower level than an item S if and only if X is a component in the mechanism for some activity ψ of S. X is a

---

[3] See also Gillett (2016), who argues that the existence of compositional relations between wholes and their constituents leaves open the issue of reductionism versus emergence. Gillett views mechanistic explanation as a kind of compositional explanation, which is a view that we reject.

component in a mechanism if and only if it is one of the entities or activities organized such that S ψs.

Consider, for example, a dividing cell. In this case, the chromosomes and other entities that together compose the mechanism responsible for cell division are at a lower level than the cell. The activities of the components of the mechanism are in turn phenomena that can be accounted for in terms of lower-level components, and so on. We have thus a hierarchy of mechanisms that grounds a level-relation among all the components of mechanisms in a given hierarchy. These are what Craver calls 'levels of mechanisms'.

On this picture, therefore, the concepts of mechanism and levels are interrelated, in the sense that the same relation both underlies the level relation and is used to understand what a (constitutive) mechanism is. In contrast, on the picture sketched earlier, levels of composition are not viewed in terms of mechanistic levels. On the account we want to defend, in particular, levels and mechanisms are distinct notions, in the sense that there is no relation that underlies them both. This means that we do not think that a relation of constitutive relevance is needed in order to understand what a mechanism is as a concept-in-use; causation in terms of difference-making is enough.

We do not of course deny that there exist relations of composition. What we deny is that in order to understand what a mechanism is in the life sciences, we need to give an account of what this composition relation consists in. What we want to claim is the following: the typical sense of mechanism in the life sciences is Causal Mechanism, and the typical sense of levels is levels of composition. Thus, in order to understand what levels and what mechanisms are in the life sciences, we should keep these notions apart.

To further see the difference between levels of composition in our sense and levels of mechanisms, consider the following strange consequence of Craver and Bechtel's account: on that view a protein is not necessarily at a lower level than a cell. If entity X is not a component in a mechanism for some activity ψ of S, it is not at a lower level than S. But a particular protein may not be a component in the mechanism for some activity of a particular cell. Also, this account does not provide an answer about whether two adjacent cells, which are not components of the same mechanism, are at the same level or not. We find both of these consequences counterintuitive.[4] When biologists talk about the cellular level, for example, what they mean is the level of organisation that concerns cells and their activities, which is above the genetic level and below the anatomical

---

[4] Craver is aware of these consequences of his view; see Craver (2007a, 192–3).

Figure 9.1   A causal pathway with a single level of composition.

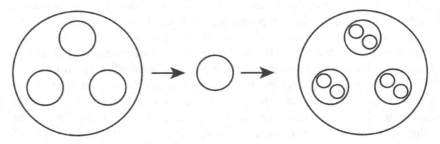

Figure 9.2   A causal pathway with multiple levels of composition.

level, for instance. In this picture, all cells, even if they are not part of the same mechanisms or wholes (because, e.g., some cells compose the muscle tissue, and other cells compose the neural tissue), are nevertheless at the same level.

We agree with new mechanists that many explanations in life sciences are multilevel. But since for us mechanistic constitution cannot be used to ground a hierarchy of levels of composition, we need now to explain how it is possible to have explanations spanning multiple levels if we accept Causal Mechanism.

9.4   Levels of Composition and Multilevel Mechanistic Explanation

How can we understand multilevel mechanistic explanation on our view? The key idea is the following: causal pathways may contain components from one level of composition only, but they may also contain components from multiple levels of composition (see Figures 9.1 and 9.2). In saying this we do not mean that causal relata in pathways are objects; components of mechanisms qua causal pathways and the entities that stand in compositional relations are different kinds of 'things'. Thus, components in a pathway are things that can stand in causal relations, for example, events or whatever one takes to be the relata of causal relations, and so typically not entities. When a pathway is represented using variables, we talk about causal relations among variables that are taken to correspond to properties of various entities. In contrast, it is typically entities that stand in compositional relations.⁵ However, the causally relevant properties in a pathway are possessed by entities that can be at various compositional levels (e.g., membrane potential is a property of the membrane, which is composed of phospholipids and other molecules), and so we will say that in this case a pathway contains entities from various levels of composition.

We will say that whatever causally contributes to the phenomenon at hand is *part of the same pathway*, and in this sense it can be viewed as being at the same explanatory level, irrespective of its level of composition. To be at the same explanatory level is to be part of the same causal pathway. Explanatory levels are to be contrasted with compositional levels. Compositional levels are formed out of compositional or mereological relations. Explanations that describe causal pathways that contain entities from many compositional levels can then be viewed as multilevel explanations. Of course, the notion of 'level' in such multilevel explanations is the compositional notion; explanatorily, as we said, components of a pathway are at the same 'level'. But since, as we saw earlier, the language of levels in biology commonly refers to levels of composition or organisation, it makes sense to emphasise *compositional* levels and call these explanations '*multilevel*'.

A crucial point here is that the issue of how exactly to understand composition– that is, what it is for a set of entities to compose a whole, and what ontological view best accounts for compositional relations – does not matter for explaining how a phenomenon is brought about. This means that one can have an account of mechanisms as causal pathways and of multilevel explanation in our sense without being committed to reductionism or anti-reductionism. The reason is that causation in Causal Mechanism is viewed in terms of difference-making: *something is a component in a pathway if it makes a difference to the effect produced by the pathway*. Now, a whole can be viewed in reductionist terms, as not having any causal powers over and above the causal powers of its organised component parts; alternatively, it can be viewed in anti-reductionist terms as possessing causal powers over and above the causal powers of its organised components parts.⁶ Either way, the whole (or the whole's properties) can act as a difference-maker and can thus be a component

---

⁵ This is not accepted by all. Gillett (2016), for example, thinks that not only entities, but properties, powers and processes can stand in compositional relations too. In Craver's picture 'acting entities', that is, a mechanism's components, can stand in compositional relations to a higher-level acting entity, that is, the phenomenon for which the constitutive mechanism is responsible.

⁶ This is just one way to view the difference between ontological reductionism and anti-reductionism (cf. Gillett 2016). It does not matter, for our argument, how exactly one understands the

in a causal pathway. We can thus understand its causal role in a pathway without being committed to a specific view about the ontology of composition. As in the case of the metaphysics of causation, this is another issue where a proponent of Causal Mechanism can remain agnostic.[7]

Let us now illustrate the idea that we can have causal pathways containing entities from several levels of ·composition by considering various examples. This will also establish that we find many multilevel mechanistic explanations in the life sciences in the sense recommended by CM.

9.5   Examples of Multilevel Mechanisms

We start with an example we have examined in detail, that is, the mechanism of apoptosis. We have seen in Chapter 3 that Kerr et al. described this mechanism at the cytological level. As they stressed, at the time it was not known which factors trigger apoptosis and what kind of cellular mechanisms are active before the morphological changes associated with apoptosis can be observed. The biochemical descriptions of the signalling pathways of apoptosis are exactly these cellular mechanisms that act as the trigger for the process described by Kerr et al.

What is the relationship between the pathway described at the biochemical level and the pathway described in cytological terms by Kerr et al.? The answer is that the biochemical pathway is a cause of the morphological changes. We could then combine the extrinsic pathway, for example, and the cytological mechanism as follows:

> Fas ligand binds to Fas receptor → adaptor protein binds to Fas receptor → procaspase-8 or 10 binds to adaptor protein → formation of DISC → activation of caspases-8 or 10 → caspases-8 or 10 activate effector caspases → destruction of proteins → condensation of nucleus and cytoplasm → budding → formation of apoptotic bodies → apoptotic bodies are phagocytosed.

---

disagreement between ontological reductionists and their opponents. The important point is that a whole can act as a difference-maker irrespective of the correct ontological account of composition.

[7] Agnosticism about the nature of composition is an important difference between Causal Mechanism and an account such as Craver's. In Craver's account, as we saw, the notions of a 'constitutive' mechanism and a multilevel mechanistic explanation both depend on the notion of constitutive relevance. Thus, an account of constitutive relevance is required. For us, such an account is not required and we can remain agnostic. One can here say that mutual manipulability is not an account about the nature of constitutive relevance, but is to be viewed in epistemic terms; but then the account seems incomplete. See also Glennan (2021, 11441), who argues that since 'objects are counted among the components of mechanisms . . . an account of corporeal composition is required to properly elucidate mechanistic constitution'.

Here we have a causal pathway containing entities from various levels of composition. Part of the pathway is described in biochemical terms and thus refers to entities that belong to the biochemical level; and part of the pathway is described in cytological terms and thus refers to higher-level structures like the nucleus and the apoptotic bodies. This, then, is a case of a multilevel mechanism.

Another example of a multilevel mechanism is the pathway of visual transduction. When light enters the eye, it is focused by a lens on the retina. Light then activates rhodopsin, a receptor protein located in rod cells, which then activates G-proteins. G-proteins, in turn, activate PDE6 (a protein complex) which catalyses the conversion of cyclic GMP to GMP. This leads to the closing of sodium channels, the cell hyperpolarises and the voltage-gated calcium channels close. As a result, glutamate release drops, which leads to the depolarisation of on-centre bipolar cells. This activates ganglion cells that activate the optic nerve, which results in the receiving of the signal by the brain (cf. Tortora & Derrickson 2012). Here, we have a pathway that contains entities from various levels of organisation. It includes photons, the lens, various proteins, protein complexes, channels and cell membranes (since hyperpolarisation is a property of the cell membrane).

Our account of the visual transduction pathway can be contrasted with how Craver and Bechtel (2007) view this example, that is, in terms of levels of mechanisms, distinguishing between processes at higher and at lower levels. The transduction of light into 'a pattern of neural activities in the optic nerve' (p. 549) is the process at the highest level. This process is then decomposed in lower-level entities and their activities. The changes in rods and cones are components in a lower-level process, and the components that explain rod activation are at an even lower level. As they put it, '[e]ach new decomposition of a mechanism into its component parts reveals another lower-level mechanism until the mechanism bottoms out in items for which mechanistic decomposition is no longer possible' (p. 549). We think that our construal of the pathway is simpler, in that it shows how a process (i.e., visual transduction) is decomposed into sub-processes, without adopting the constitutive account of mechanism (and levels of mechanisms), which leads to positing non-causal constitutive relations between 'acting entities' (in Craver's sense) at different levels. For us, what decomposition amounts to is finding intermediate causal steps between an initial cause (e.g., the entering of light into the eye) and an effect (e.g., the activation of the optic nerve).

In Chapter 4 we examined the case of scurvy and we saw that scientists discovered the following causal pathway:

Citrus fruits → vitamin C → scurvy.

This is a very simplified description of the causal pathway, but the important point for our purposes is that this is again a multilevel causal pathway, as it contains entities from several levels of organisation. The outcome of the pathway in particular, that is, the development of scurvy, is something that concerns the whole organism, as the symptoms of scurvy include weakness, feeling tired, sore arms and legs, which are properties of the organism as a whole. Similarly, the dietary habits of the organism, in particular the presence or not of citrus fruits in the diet, is again something that has to do with the whole organism. In contrast, the mediator, that is, presence or not of vitamin C, concerns a lower level of organisation. A more complete description of the pathway will have to explain how lack of vitamin C disrupts various biosynthetic pathways such as the synthesis of collagen, dopamine, epinephrine and carnitine; these pathways concern the biochemical level of organisation. It will also have to mention how lack of vitamin C affects various tissues, such as skin, gums and bones, which concern a higher level of organisation. Thus, the pathway that leads to scurvy contains entities from several levels of organisation, as it describes entities at the biochemical level, at the level of tissues and at the level of the whole organism.

Like the pathway of scurvy, many pathological mechanisms include entities from various levels of organisation. A description of a pathological mechanism may mention entities at the levels of genes, biochemical pathways, cells, tissues, behaviour of organs and properties of the whole organism. Causes of a disease, in particular, include environmental factors like radiation and temperature extremes. Let us look at an example that exhibits many of these features: the mechanism of development of type 2 diabetes.

Diabetes is a syndrome characterised by hyperglycaemia (high blood sugar) due to deficiency of insulin (Bugianesi et al. 2005; Gardner & Shoback 2017). Insulin is one of the main hormones that regulate glucose homeostasis by stimulating glucose uptake. Main symptoms of type 2 diabetes are increased thirst and hunger, frequent urination, weight loss and feeling tired. Development of type 2 diabetes is due to insulin resistance (which means that cells do not respond normally to insulin) and due to a defect in pancreatic β-cells. Pancreatic β-cells release insulin as a response to increased levels in glucose in the blood. Normally, β-cells can compensate for insulin resistance by increasing insulin production, but defective β-cells cannot. This failure of β-cells is mainly due to underlying genetic factors. So, in type 2 diabetes glucose homeostasis is impaired. As diabetes progresses, various mechanisms may cause the function of β-cells to decline further, resulting in worsening hyperglycaemia.

Type 2 diabetes has a 'natural history' that progresses from an early stage with insulin resistance but no symptoms to a stage with mild hyperglycaemia and finally to a stage where pharmacological intervention is required. This natural history is roughly as follows. The development of insulin resistance is influenced by both genetic and environmental factors such as obesity and physical inactivity. There is an initial period before the development of type 2 diabetes, where insulin resistance leads to hyperinsulinemia, which means that the pancreatic β-cell can produce high levels of insulin and overcome insulin resistance, so that normal glucose homeostasis is maintained. But eventually defective β-cells in combination with insulin resistance cannot maintain glucose homeostasis; at this point, type 2 diabetes is diagnosed and can lead to microvascular and cardiovascular complications (e.g., diabetic kidney disease, retinopathy, coronary artery disease, stroke).

The pathophysiology of the disease includes three main defects, which are typically described at the level of organs and tissues and involve factors such as levels of glucose and insulin. These defects concern the activities of the pancreas, the liver, and the muscle and adipose tissues. The first defect concerns the pancreatic β-cells that, as we have seen, cannot produce enough insulin. The second defect is that due to the decrease in insulin and insulin resistance, glucose uptake at muscle and adipose tissues is decreased. The third defect is an increase in glucose production in the liver, which is normally regulated by insulin; due to the decrease in insulin, hepatic glucose overproduction can no longer be restrained. All three of these defects result in hyperglycaemia.

Many of the molecular mechanisms involved in type 2 diabetes are known. For example, the action of insulin in stimulating glucose uptake can briefly be described as follows: insulin binds to its receptor and activates a pathway that eventually allows glucose to enter the cell. The binding of insulin to its receptor located in the cell membrane causes tyrosine phosphorylation of insulin receptor substrate (IRS) proteins, which eventually causes glucose transporter 4 to be translocated to the cell membrane, where it allows glucose to enter the cell. Various defects in this pathway can lead to insulin resistance. In the muscle cell, for example, the cause of insulin resistance is serine rather than tyrosine phosphorylation of IRS proteins. In such a case, glucose is prevented from entering the cell.

The mechanism of development of type 2 diabetes, then, involves many organs and functions, many interrelated physiological and molecular mechanisms, is described at various levels of organisation and is characterised by a natural history.

Evolutionary mechanisms provide more examples of mechanisms that involve entities from various levels of composition. Evolutionary mechanisms such as natural selection have presented problems for new mechanists (as we will see in Chapter 10), since it is not evident what exactly counts as entities, activities and organisation in these mechanisms. Mechanism talk, however, is very common in evolutionary biology, so the question arises in what sense processes like natural selection and genetic drift are mechanisms. We think that CM provides an easy answer; the main reason, we think, to say that these processes are mechanisms is that they identify the causal steps of how evolutionary change comes about (we will come back to this point in Chapter 10). The important thing to note here is that descriptions of processes of natural selection can include genes and their frequencies in populations, traits of organisms, properties of the environment as well as properties of developmental systems, such as robustness, phenotypic plasticity and modularity, that is, entities from several levels of organisation.

As a last case, let us also briefly discuss a main example that Craver (2007a) examines: the multilevel mechanism of spatial memory. Craver takes this mechanism as having four main levels. The level of spatial memory is at the highest level; then, we have the level of spatial map formation, the cellular-electrophysiological level (which includes the mechanism of long-term potentiation) and, last, the molecular level that includes the molecular mechanisms (e.g., the function of NMDA receptors that underlie the chemical and electrical properties of nerve cells). Craver's general argument is that these four levels of spatial memory are more appropriately viewed in terms of levels of mechanism, and not in terms of other notions of levels, such as levels of size or of spatial containment. Craver, then, takes the mechanism of spatial memory to 'include NMDA receptors as components in LTP mechanisms, LTP as a component in a hippocampal spatial map mechanism, and spatial map formation as a component in a spatial memory mechanism' (p. 266), and he thinks experiments provide powerful evidence that 'the phenomena at each of these levels – NMDA receptor function, LTP, spatial map formation, and spatial memory – is constitutively relevant to the next' (p. 265).

For us, the relevance relations among all these components are causal relations: the function of NMDA receptors is causally relevant to the operation of LTP mechanisms, which are parts of the spatial map mechanism. When we go up a level, we consider a more extended causal network. The mechanism of spatial memory, the 'highest level' in Craver's account, can be viewed as the most extended causal network, which includes NDMA receptors as components. When we have such a network we can zoom in all the way to the molecular level, that is, to a size scale where we can identify the molecular details of the mechanism, or zoom out to the scale of neurons or brain regions.

Moreover, in the extended causal network we have entities from several levels of composition. This is important for making sense of interlevel experiments. For example, Craver mentions a bottom-up interlevel experiment where researchers knock out a gene that encodes a subunit of the NMDA receptor, so that knockout mice do not have functional NMDA receptors, and test their performance in the Morris water maze; the result is that knockout mice perform far worse than controls. Change in the performance of the knockout mice, which we take as the causal outcome of the bottom-up change, is a change in the behaviour of an entity at a higher level of organisation than the NMDA receptors, that is, an organism as a whole. Multilevel mechanisms containing components from various levels of composition, then, can account for interlevel experiments.[8]

9.6 Mechanisms and Interlevel Causation

We take the examples of mechanisms discussed in Section 9.5 to illustrate a main point of this chapter, that is, that mechanisms that include components from several levels of composition are very common. Scientists, then, commonly make causal claims that involve interlevel causes, which, however, are often viewed with suspicion by philosophers. In this last section we will argue that interlevel causation is unproblematic, contrasting again our account with Craver and Bechtel's analysis.

A difficulty of accepting interlevel causation is that things related by mereological relations are not distinct in a way that would enable them to stand in causal relations. This consideration is not a problem for the view that there exist multilevel causal pathways in our sense, since in several of the cases we have examined the components of the pathway do not stand

---

[8] For a critical examination of Craver's example of spatial memory, see also Eronen (2015). Eronen adopts what he calls a 'deflationary approach' to levels; he suggests that we should prefer more well-defined concepts, especially when we think about causation, such as scale and composition.

in mereological or compositional relations to each other. For example, the apoptosome is composed of several proteins, and is thus at a higher level of composition than, for example, an Apaf-1 protein (which is one of its components). But there is no difficulty in saying that the apoptosome is a component in a causal pathway that also includes entities at a lower level of composition as components. Such causal claims are unproblematic, as are claims that smaller entities can cause changes in bigger ones, and vice versa (however, in cases such as the example of type 2 diabetes, components of molecular pathways are spatiotemporally contained within the individual that develops type 2 diabetes; we will discuss this case below).

However, why posit an interlevel causal relation at all? Consider a causal pathway such as the case of scurvy. Here, the cause is at a higher level, since it concerns dietary habits, while the presence or absence of vitamin C concerns a lower level of organisation. But then one can argue as follows. When we have a case of such interlevel causal claims, what really happens is that the lower-level constituents of the putative higher-level cause do the real causal work, and thus we do not need to posit an interlevel causal relation. According to Craver and Bechtel, for example, there is no inter-level causation. They view cases that seem to involve interlevel causal relations either as cases of constitutive relevance, if the putative causal relation is between a component and the mechanism as a whole, or in terms of what they call 'mechanistically mediated effects'. These are 'hybrids of constitutive and causal relations in a mechanism, where the constitutive relations are interlevel, and the causal relations are exclusively intralevel' (2007, 547).

Take, for example, the claim that infection with a virus led to the death of the general; here, according to Craver and Bechtel, we have a causal claim that concerns how the virus interferes with various mechanisms in the organism, ultimately producing 'the physiological conditions that constitute the general's death' (p. 557). Similarly, in cases of putative top-down causes, the top-down cause is constituted by some mechanism, which then produces an outcome. This is why, then, cases of putative interlevel causation are described as 'hybrids', since 'the putative interlevel claim is analyzed into a causal claim coupled with one or more constituency claims' (p. 561).[9]

---

[9] Bechtel (2017) argues that the account in Craver and Bechtel (2007) in effect renders higher levels epiphenomenal, as it 'suggests a highly reductionistic picture of levels according to which causal relations that were supposed to be between entities at higher levels of organization dissolve into causal interactions at the lowest level considered' (p. 262).

We do not think that we need to adopt a hybrid picture such as the one suggested by Craver and Bechtel in order to make sense of interlevel causal relations in multilevel causal pathways. First, as we have argued in Chapter 8, we reject the constitutive account of mechanism that is presupposed in Craver and Bechtel's view. We think that mechanism as a concept-in-use in the life sciences is captured by CM and so it does not incorporate any compositional relations, just causation as difference-making. Second, we think that interlevel causal claims in science should be taken at face value and that we should try to develop a philosophical account that is adequate to capture these claims, rather than dismissing a literal construal of interlevel causal claims in part because of philosophical intuitions such as that wholes cannot causally influence their parts and vice versa. To return to the example of the mechanism of development of type 2 diabetes, to say that a defect in β-cells and insulin resistance cause type 2 diabetes or that type 2 diabetes can result in cardiovascular complications is to make causal claims (thus taking scientific talk of interlevel causation literally) and not to say something about the molecular mechanisms that underlie the natural history of type 2 diabetes and constitute the higher levels of mechanisms or organisation involved in this case. Third, to adopt interlevel causal claims where the causal relata involve properties of entities that are themselves (i.e., the entities) related as part and whole is not to say that parts cause wholes or vice versa in a synchronic, and thus problematic, manner. It is important here to take into account the temporal dimension, where, for example, defects in the insulin pathway and β-cells over time lead to changes that concern higher levels of organisation, irrespective of the fact that insulin pathways and β-cells are spatiotemporally contained in the organism.

We thus take interlevel causal relations to be conceptually coherent; moreover, we take it that interlevel causal claims are very often to be preferred to causal claims that involve only lower levels of organisation – for example, trying to couch all mechanistic causal explanations in molecular terms. This is because, as Gilbert put it, each level of organisation is in some sense irreducible to those below it. This can be made more precise by using Woodward's notion of conditional independence that we think captures part of what Gilbert means. Woodward (2020) defines conditional independence as follows. Suppose that we have a set of variables L that are causally related to E, but that we can use higher-level variables U that 'correspond to a coarsening of the L variables' and can be used to 'summarize the impact of the L variables on E'. This means that conditional on the values of U, further details about L won't matter – U 'screen off' L from E. Variables L in this case 'are *independent* of E, *conditional* on U' (p. 428). As Woodward puts it, 'on this view of the matter, claims of downward causation (and claims of interlevel causation more generally) can be thought of as claims about the irrelevance of certain kinds of information conditional on other sorts of information – we can legitimately make claims of interlevel causation when such conditional irrelevance relations are present' (p. 444). For example, when we say, in the case of the Hodgkin-Huxley model, that the membrane potential V is a cause of the ionic currents and channel conductances, 'any further information about how that potential is realized in the electromagnetic forces associated with individual atoms and molecules does not matter' (p. 444) for the causal impact that V has on the other variables. Woodward's account then shows when we have reason to appeal to interlevel causation.[10]

In sum, then, causal pathways can be multilevel in the sense that they can contain entities from various levels of organisation. Interlevel causal relations are not problematic and, as we have seen, are ubiquitous in the life sciences. This account of multilevel mechanistic explanation is simpler than Craver's account, which relies on the notion of levels of mechanisms, as well as Craver and Bechtel's hybrid picture of interlevel causation, as we do not use any constitutive relations to give an account of what a mechanism is; the notion of a mechanism and the notion of levels (of composition) are for us distinct notions. Last, this account of multilevel mechanistic explanation can remain agnostic regarding the issue of how exactly composition is to be understood.

---

[10] Apart from conditional independence, another reason to appeal to interlevel causation, as Woodward notes, is that explanation in terms of lower-level causes is often computationally and epistemically intractable. It is important to note also that Woodward frames his analysis in terms of what he calls an 'interactionist' notion of levels that he takes to be conceptually distinct from the notion of levels of composition or size.

CHAPTER 10

Methodological Mechanism

10.1 Preliminaries

In the previous chapters we have defended what we have called *Causal Mechanism*, that is, the view that mechanisms (especially in the life sciences) are causal pathways that are described in theoretical language, where the pathway is underpinned by networks of difference-making relations. We have also characterised CM as metaphysically agnostic. This is an especially important feature of CM, since it differentiates it from what we have been calling inflationary accounts of mechanism, which we take all major accounts to be. In Chapters 8 and 9 we also argued that non-causal constitutive relations are not required to understand what a mechanism is in biological practice. CM is best seen in the context of a thesis that we call, following Woodger (1929) and Brandon (1984), *Methodological Mechanism* (MM). It will be the main aim of this chapter to develop and argue for MM as a general framework for understanding the search for mechanisms.

10.2 Methodological Mechanism: Historical Predecessors

10.2.1 Woodger on 'Methodological Mechanism'

Woodger (1929) distinguished between two ways in which a certain notion can be employed: a metaphysical or ontological way and a methodological one. The latter is when a notion is used for the purposes of description 'independently of its metaphysical interpretation'. In this case, Woodger says, the notion 'is employed methodologically, that is, simply for the purpose of investigation' (p. 31). The advantage of this use is that the notion can be used in a certain practice and cast light on it *independently* of whatever difficulties (and controversies) are raised by the intricate metaphysical debates concerning what its worldly reference is really like. In his discussion of mechanism, Woodger says that the notion of mechanism can be employed in precisely this methodological way, independently of how it is metaphysically interpreted.[1]

As a justification for this view, Woodger cites embryologist Gavin de Beer, who understands the mechanistic viewpoint in biology as follows:

> [The mechanistic point of view] in no way commits one to the 'materialistic' idea of life. Neither does it mean that life is nothing but physics and chemistry. What this point of view does stand for is that whatever the processes of life may be, they work in an orderly way, producing similar effects under similar conditions. *Steering between 'materialism' and 'vitalism', this conception has become known as mechanistic.* (quoted in Woodger 1929, 258; emphasis added)

For de Beer, then, the mechanistic viewpoint is not an ontological thesis, but concerns the nature of the object of study that must be presupposed for (mechanistic) science to be possible. In contrast to materialism and vitalism, it affirms no specific ontology for biology. Here is Woodger's own reading of this:

> It seems clear from this that all that is here meant by mechanism is the belief in the 'law of causation' or the 'uniformity of nature'. This is commonly regarded as a necessary methodological postulate of natural science.... Psychologists for example, speak of the 'mechanism of a neurosis' referring thereby I suppose to the ordered 'structure' of psychical processes. They assume that there is some such orderly structure and call it a mechanism without implying anything further about the ontological nature of those processes. *Thus the term may be used for the methodological postulate that there is some sort of order, and then it may be applied to that order itself.* (emphasis added)

He explains further:

> [For] de Beer, the term [mechanism] was used for a methodological postulate which asserts that the object of study is in some way orderly, and we saw how this term may then be extended to that which is thus assumed to be orderly but without any further metaphysical assumptions about its 'nature'. (pp. 259–60)

A 'mechanism' as a concept-in-use, then, according to Woodger's construal of de Beer's view, can be taken to mean an ordered causal

---

[1] Woodger (1929, chapter 5) distinguishes between four different senses of mechanism. These are (1) mechanism in the sense of classical mechanics, (2) mechanism as an explanation that uses only concepts from physics and chemistry, (3) mechanism in the sense of an analogy to a machine and (4) mechanism as a kind of a methodological postulate.

structure that scientists discover and that they describe in theoretical terms, without any further specification as to the fundamental nature of this order. The mechanistic point of view amounts to the methodological thesis that the aim of science is to discover such ordered causal structures. In the account developed in this book, instead of de Beer's ordered structures, we have causal pathways responsible for the phenomena. We can retain the idea of an ordered causal structure (except in indeterministic cases), adding that what makes a pathway *causal* are the difference-making relations among the components of the pathway.

According to Woodger, we need to 'ask the methodological mechanist what he has to say in support of his contention that the mechanical explanation is the only one which is admissible in science' (p. 231). Hence, Methodological Mechanism is a view about mechanistic explanation and its admissibility, and not about the blueprint of the universe. It's not about the metaphysics of mechanism, but about the use of the concept of mechanism in science and in particular about the importance of identifying causal pathways. In adopting this view, Woodger noted that mechanism is a 'methodological postulate' which as such 'makes no assertions about the nature of the processes studied, but merely asserts that they take place according to law, or "work in an orderly way"' (p. 258).

Taking a cue from Woodger's Methodological Mechanism, we want to claim that commitment to mechanism in science is adopting a *methodological postulate* which licenses looking for the causal pathways for the phenomena of interest. Hence, MM licenses adopting Causal Mechanism. CM, as we saw, allows for a rich understanding of the use of this concept in biology (and other sciences) without getting embroiled in a debate about what things in the world mechanisms *really* are and what kind of metaphysical categories their (theory-described) components fall into. Viewing mechanism as a methodological thesis allows that the sought-after identification of the causal pathway by which a specific result is produced is fully captured in the language of the specific theory, using deeply theory-laden concepts. It forfeits any further need, for the purposes of understanding how mechanisms explain, to offer a general metaphysical account of how the theory-described entities and processes – the *causal pathway* – fall into neat metaphysical categories.

It bears stressing that the key feature of MM is that it is non-committal about fundamental ontology. It adopts the postulate that scientists should always try to identify the way that a particular phenomenon is produced, but it says very little about how causation itself is to be understood: it asserts only what is required for making sense of the practice of looking for mechanisms. Hence, MM is philosophically neutral. But this does not mean that MM is scientifically neutral. Insofar as it is adopted it licenses mechanistic (and only mechanistic) explanations of the phenomena or the behaviours to be explained.

Though MM does not commit us to a specific view about how causation is to be understood from a metaphysical point of view (e.g., it need not commit itself to the view that interaction is the transmission of conserved quantities, etc.), MM can still clarify the close relations between causation, explanation and (the identification of) mechanisms: at least when there is no genuine indeterminism, whatever happens has a prior cause and identifying the way the cause brings about the effect is identifying the causal pathway by means of which the cause operates.

MM, we will argue below, illuminates practice in a way that ontologically inflated accounts of mechanism do not. It accounts for the centrality of mechanisms in scientific discovery and explanation, since according to it, discovering mechanisms (i.e., causal pathways) is the central task of science. At the same time, however, it refrains from imposing on scientific practice ontic constraints that are not licensed by it. According to MM, the mechanistic view need not be taken as something stronger than a certain methodological commitment to a kind of explanation.[2]

10.2.2   Brandon on 'Mechanism'

Brandon (1984) also argues for a methodological position he calls mechanism. He claims that mechanism, in this sense, should not be identified with reductionism, and that '[b]iological methodology is thoroughly mechanistic' (p. 348). He takes this position to have been described by Marjorie Grene (1971), when she writes:

> [L]et us look for a mechanism which might underlie the phenomena we hope to understand, seeking wherever we may relevant sources from which

---

[2] Consider also the following quotation by the zoologist Lancelot Hogben (as given by Brandon 1984), who views mechanism as a primarily epistemological view: '[I]n any discussion between the two [mechanist and holist or vitalist], the combatants are generally at cross purposes. The mechanist is primarily concerned with an epistemological issue. His critic has always an ontological axe to grind. The mechanist is concerned with how to proceed to a construction which will represent as much about the universe as human beings with their limited range of receptor organs can agree to accept. The vitalist or holist has an incorrigible urge to get behind the limitations of our receptor organs and discover what the universe is really like' (Hogben 1930, 100).

to derive, first, an analogue of a possible mechanism, and then, if we are shrewd and lucky and experience bears us out, maybe a description of the mechanism itself. (pp. 63–4)[3]

For Grene and Brandon, then, the central task of science is to search for mechanisms that produce the phenomena. This is the methodological position that Brandon calls 'mechanism'.[4] This leads to a mechanistic explanation that 'tells us how in fact those phenomena are produced' (Grene 1971, 64). But what is here meant by a mechanism? Here is Brandon again:

Here I cannot be precise. Sometimes old-fashion spring-wound clocks and watches are called mechanical devices, in contrast modern battery-powered digital watches are called electronic devices. Clearly, I cannot use 'mechanism' in such a narrow sense. Mechanisms may consist of springs and gears, they may consist of computer chips and electrical pulses, they may consist of small peripheral populations and geographic isolating barriers. I cannot delimit all possible mechanisms because it is the business of science to discover the mechanisms of nature. At best I could list the sorts of mechanisms science, or more specifically, biology has discovered.... To model a process is to offer a more or less plausible hypothesis concerning the mechanism underlying the process. *Thus any process capable of being modelled is a mechanistic process.* (1984, 346, emphasis added)

Brandon suggests that the opposition between mechanism and various non-reductionistic ontologies, such as vitalism and holism, rests on the mistake that 'mechanistic methodology has been seen as implying (or somehow supporting) a reductionistic ontology' (p. 347). Brandon thinks, following Grene, that mechanism supports a multilevel ontology (Grene calls this 'level-pluralism').

Brandon (1990, 185) returns to the question 'What is a mechanism?' He says:

A causal/mechanical explanation is one that explains the phenomenon of interest in terms of the mechanisms that produced the phenomenon. What is a mechanism?... [T]his question has no general metaphysical answer, because the business of science is the discovery of mechanisms; so we cannot delimit in any a priori manner the mechanisms of nature.... The best we can do is to give an open-ended answer: *a mechanism is any describable causal process.* (emphasis added)

---

[3] Grene follows Harré, saying that 'the central task of science in Harrean terms is the imaginative construction of theoretical models which suggest ways in which particular sets of phenomena may be produced'. For Harré's views on mechanism, see Chapter 5.
[4] Grene does not use 'mechanism' as a label for this methodological position.

We think that Brandon's position here can be generalised as follows: concepts such as *mechanism* that are central in scientific practice should be viewed as methodological postulates rather than as presupposing robust metaphysical commitments. But methodological postulates should be 'open-ended'; otherwise, they would unnecessarily limit research. This, we think, does not render MM a trivial thesis (we will return to the triviality objection in Section 10.6). Far from being a trivial commitment, MM is flexible enough to foster searching for mechanisms, whatever the ontic signature of the world might be.

Consider again Woodger's question: What has the methodological mechanist to say 'in support of his contention that the mechanical explanation is the only one which is admissible in science'? Although we do not think that mechanistic explanation is the only admissible form of explanation in biology and elsewhere, we agree with Woodger that this question should be answered.[5] We agree with Grene and Brandon, as well with the more recent mechanists, that the search for mechanisms is an important aim in science in general, and in life sciences in particular. For many new mechanists, however, this methodological norm is understood in terms of inflationary metaphysics. In what follows, we will argue that such accounts weaken the normative force of the main mechanistic methodological norm. We will argue that for the methodological norm to have its full force, mechanism should be understood along the lines of CM. We take this argument to constitute an answer to Woodger's question, in the sense that the argument will illuminate why the main mechanistic methodological norm has the place it has in current science. So, this argument will be a further reason to accept CM.

10.3 General Characterisations of Mechanism

10.3.1 Descriptive Adequacy

We take it that any general concept of mechanism adequate to the aims of many new mechanists has to satisfy at least two important adequacy conditions. First, the concept we seek to clarify must be central in scientific practice; second, it should be common across scientific fields. The first

---

[5] Here is Brandon's answer to this: 'I've argued that biological methodology is thoroughly mechanistic, but why should it be? As discussed in the first section of this paper, Grene has argued that mechanistic explanations give, or attempt to give, one an understanding of how the phenomena to be explained are really produced. I have nothing more to add to that discussion except to baldly state that that is what science, or more specifically, biology, ought to be up to. Thus, on this view not only is biological methodology mechanistic, it ought to be' (1984, 350).

condition stems from New Mechanism's central aim to give an account of science that is as close as possible to actual scientific practice. It says that the general concept has to feature in scientific practice and be in conformity with how scientists themselves use the concept, that is, to be a *concept-in-use*, as we like to call it. The second condition is justified given New Mechanism's aspiration to be a framework to be applied not only to a particular area within science, but to life sciences in general, as well as to many other scientific disciplines, from the social sciences to physics. The hypothesis that there exists such a concept common across diverse fields is explicitly endorsed by many new mechanists (see Illari & Williamson 2012; Glennan 2017).

To further appreciate the importance of this commonality condition, consider the trend in recent mechanistic literature to characterise in general terms various specific instances of mechanisms found in particular scientific fields (see, e.g., Glennan & Illari 2018a, chapters in part 4). Such more local accounts of mechanism cannot remove the need for searching for a characterisation of the general concept, even if there are more specific uses of this concept in particular cases. Clearly, one can always raise the question: *in virtue of what* are all those different species of mechanisms members of the same genus? What makes them all mechanisms? But then, a general characterisation of mechanism is still needed; such a characterisation would in that case be an explication of the general concept of which the various more particular kinds of mechanisms are instances.[6]

10.3.2 Normative Adequacy

General characterisations of mechanism have also to conform to a kind of normative adequacy condition. If we take mechanism to be a concept that is really present in actual scientific practice, then it has to have an influence in directing or regulating practice.[7] This regulating role can best be seen in

---

[6] Of course, a more radical pluralist stance is also available: perhaps there is no overarching concept of mechanism, but various distinct notions that have to be distinguished. Such a view, however, would undermine both the main working hypothesis of many new mechanists that there exists a general concept and, more importantly, the significance of the methodological and ontological theses of New Mechanism (see Sections 10.3 and 10.4).

[7] Our approach shares similarities with what Woodward (2015) calls a 'functional' project, which he contrasts with a metaphysical project, among others. He explains what a functional project is in the case of causation as follows: 'by a functional approach to causation, I have in mind an approach that takes as its point of departure the idea that causal information and reasoning are sometimes useful or functional in the sense of serving various goals and purposes that we have. It then proceeds by trying to understand and evaluate various forms of causal cognition in terms of how well they conduce to the achievement of these purposes' (pp. 693–4). In contrast to the metaphysical and other projects, the functional project has a 'normative or methodological dimension' (p. 694). Our examination of

the context of what we can call the *Methodological Tenet*, which can be formulated as follows:

(MT) Scientists, in investigating the phenomena, should search for mechanisms responsible for them.

MT has to be viewed in the context of the condition of descriptive adequacy mentioned earlier: since mechanism is taken to be a central concept-in-use active in diverse scientific fields, MT must be pervasive in science. MT is best seen as a methodological norm that guides scientific practice; it says what a main aim of science should be and how scientists should proceed in investigating the phenomena.

There are two points that should be stressed concerning MT. The first is that, as formulated, MT does not specify what a mechanism is. But without unpacking the meaning of 'mechanism', it is not clear what MT amounts to; unless we do this, its informational content remains unclear and thus MT is unhelpful as a guide for research. The following (incomplete) version of MT, then, needs to be fleshed out, by inserting a specific characterisation of mechanism:

(MT*) Scientists, in investigating the phenomena, should search for mechanisms responsible for them, where a mechanism is ⟨. . .⟩.

Second, MT being a thesis about methodology, one can ask whether there exists any general argument to the effect that searching for mechanisms must be a central aim of science (this is a version of Woodger's question). The more normative force we take MT to have, the more urgent this request is.

These two points are interrelated: the completion of MT* by providing a characterisation of mechanism must not be such as to weaken the normative force of MT. In that sense, MT constrains potential candidates for a general characterisation of mechanism. This, then, is a second kind of adequacy condition to any general account of mechanism, over and above its descriptive adequacy.

10.4 Inflationary New Mechanism

Most new mechanists describe mechanisms as things with a specific ontic signature; they put emphasis on the metaphysics of mechanisms.[8] Some,

---

mechanism as a concept-in-use can be viewed as a functional project, since we are interested in the role and usefulness of this concept within scientific practice.

[8] Glennan (2017) is a representative recent example; see also Krickel (2018).

admittedly, focus primarily on epistemology and methodology rather than ontology, analysing how scientists discover mechanisms and construct mechanistic explanations (Bechtel 2006; 2008; Craver 2007a; Craver and Darden 2013). However, some version of this ontological viewpoint is implied by all dominant formulations of the concept of mechanism qua concept-in-use: as we saw in Chapters 1 and 4, the general characterisations offered are formulated in terms that are more or less ontological. New mechanists thus share the view that there is a route that leads from the philosophical elucidation of practice to substantial conclusions about the blueprint of the world.

New Mechanism, then, combines two attitudes towards mechanism as a concept-in-use. On the one hand, the concept of mechanism is taken to play a central role in practice, in discovery and in explanation. On the other hand, it is taken to tell us something about the structure of the world: mechanisms are what science discovers, and they are taken to be the building blocks of reality. Let us call this second thesis the *Ontological Tenet*. Analogously to MT* we then have:

(OT*) The world consists of mechanisms, where a mechanism is ⟨. . .⟩.

We will now argue that putting these two tenets together leads to a tension which is detrimental to New Mechanism.

10.4.1 A Central Dilemma for New Mechanism

Here is how the dilemma arises. On the one hand, New Mechanism tries to flesh out MT* and OT* by means of the general characterisation of mechanism that it abstracted from scientific practice. Since OT is offered as a substantial ontological thesis about the world, the general characterisation must have sufficient content for OT to be able to function as such. Moreover, as the underlying motivation here is that metaphysical conclusions must be directly derived from practice, there is no independent source to provide content to flesh out OT*: all such content must be provided by the general characterisation of mechanism that is grounded in scientific practice.

On the other hand, the general characterisation of mechanism must be normatively adequate. So, it must be able to guide scientific practice in a way that the usefulness of MT as a methodological norm is maximised. But then, MT should avoid being overly specific, so that instead of regulating scientific practice it ends up constraining it. Now, if mechanism as a concept-in-use were to be formulated in very specific ontological terms so as to accurately describe the building blocks of a mechanistic world and thus to provide OT* with ontological oomph, the normative force of MT would be very weak; thus, the general characterisation would fail to be normatively adequate, which is an essential adequacy condition for any general account of mechanism.

Here, then, is how we can describe the unstable nature of the combination of MT and OT in general terms: the more content the general characterisation of mechanism has so that OT can be a robust ontological thesis about the world, the more this weakens MT; and the more defensible MT is as a central methodological maxim that regulates scientific practice, the more this weakens OT. The challenge, then, is to find a general characterisation of mechanism as a concept-in-use that can simultaneously satisfy both OT and MT.

10.4.2   The Central Dilemma in More Detail

Let us assume that one wants to provide that content to OT* so that it becomes a robust thesis about the world, by taking a mechanism to be a causal process that only involves material particles in motion, as seventeenth-century corpuscularians had hypothesised – for the sake of example, let us take these particles to be characterised only in terms of a typical seventeenth-century list of mechanical properties, for example, extension, shape, size, impenetrability and motion. Surely, as an ontological tenet, this is very informative: everything that exists in the world consists in matter describable in terms of a specific list of 'mechanical' properties. But, as our argument about Newton in Chapter 1 showed, such a construal of mechanism did in fact put limits on the science of the seventeenth century. One can view the criticisms against Newton from such figures as Leibniz to arise from a commitment to the ontological constraints that should regulate scientific explanation. Newton's project can similarly be seen as involving a liberation of scientific explanation from such ontological constraints. Note here that this liberation has been fully embraced by new mechanists, who have adopted a much more liberal construal of mechanism than their seventeenth-century predecessors.

This point can be generalised: if mechanism were to be explicated in strong reductionist terms, this would surely lead to a substantial OT but to a very weak MT. Why suppose, for example, that the only legitimate mechanistic explanations in biology are explanations in terms of what happens at the molecular level? As we have seen in Chapter 9, a causal pathway can involve entities from various levels of composition; such higher-level components, then, feature in legitimate mechanistic explanations and are causally relevant and even indispensable.

But does mechanism instead support anti-reductionism? Recall that methodological mechanists are not committed to a version of materialism, which is an ontological thesis, and similarly are not committed to some form of vitalism or holism, which are also typically viewed as ontological theses. Grene and Brandon, of course, as well as most new mechanists, view mechanism as an anti-reductionist position. We agree with new mechanists that mechanism leads to explanatory anti-reductionism. But we do not think that, ontologically, the mechanist needs to cling to a firm view on this matter. This is because mechanism as a concept-in-use does not involve any commitments about whether a whole is something over and above the properties of its parts and their organisation.[9] To view properties of wholes as being causally relevant and capable of featuring in causal pathways is not a reason to opt for ontological anti-reductionism. Indeed if, as Brandon insists, mechanism should not be opposed to ontological reductionism, similarly mechanism should not be viewed as ontological anti-reductionism; for then, it would indeed be correct to oppose it to ontological reductionism too. The proper attitude for a methodological mechanist, then, should be suspension of judgement regarding this issue.

Actually, we want to claim something stronger: mechanism as concept-in-use does not (and should not) involve a commitment to any kind of ontological theses that do not seem to have a function within scientific practice.

10.4.2.1 *Inflationary Accounts Have Ontological Excess Content*

Consider again the Newtonian move against New Mechanism that we described in Chapter 1. We argued that as old mechanists like Leibniz put ontic constraints to scientific explanation, new mechanists too put ontic constraints to current mechanistic explanation. They do this by requiring that a mechanistic explanation conform to a description of mechanism given in metaphysical terms. Glennan expresses a typical view among new mechanists when he says that a mechanistic explanation shows 'how the organized activities and interactions of some set of entities cause and

---

[9] It does not matter here how exactly ontic reductionism or anti-reductionism is construed; the point is that a methodological mechanist need not have a view about this ontological issue (see Gillett (2016) for a recent discussion about how the ontic versions of reductionism and anti-reductionism should be understood).

constitute the phenomenon to be explained' (2017, 223) and that it 'always involves characterizing the activities and interactions of a mechanism's parts' (p. 223). But to say all this is to subject mechanistic explanation to ontic constraints not warranted by scientific practice.

New mechanists motivate the inflationary accounts by arguing that they are descriptively adequate to capture specific cases of mechanisms in neurobiology and molecular biology. But certainly CM, which takes mechanisms to be theoretically described *pathways*, is descriptively adequate too (more on this in Section 10.5). Moreover, we take CM to be a very clear account of what a mechanism is, which is fully grounded in scientific practice. By contrast, the inflationary accounts give rise to a series of questions about how activities relate to entities, the metaphysics of causation or the nature of the constitution relation that are heavily debated among new mechanists. As the answers to these questions are important in order to understand the content of the inflationary accounts we think it is fair to say that the inflationary accounts are not as close to practice as CM. So, descriptive adequacy cannot be the reason why inflationary accounts are to be preferred.

10.4.2.2 *Inflationary Accounts Can Constrain Practice*
We in fact think that mechanism as a concept-in-use shouldn't be inflationary, the reason being that any ontological commitments that go further than those of CM would weaken the methodological tenet of mechanism. Consider again Old Mechanism; if mechanism were to be viewed in very strong reductionist terms, this would greatly weaken the regulative force of mechanism as a concept-in-use. But mechanism as a methodological tenet is taken by new mechanists to have a central place within scientific practice. Any very strong ontological commitments would count against this central place, as they would weaken MT.

Consider, for example, the MDC account or Minimal Mechanism: What possible reason could there be for insisting that all mechanistic explanations should be in terms of organised entities and activities? Even if this requirement is much more minimal and plausible than insisting that explanations should be couched in mechanical or physicochemical terms, it nevertheless puts a constraint on practice. There is, for example, much discussion among new mechanists about whether evolutionary mechanisms such as natural selection can be captured by the dominant general characterisations of mechanism. So, Robert Skipper and Roberta Millstein (2005) have argued that the mechanism of natural selection cannot be captured by the MDC and Glennan's earlier complex system account.

According to Skipper and Millstein (2005), natural selection lacks the decomposability and organisation that (early) Glennan and MDC view as requirements for something to be a mechanism. For example, they note that it is not clear whether the environment or perhaps some parts of the environment (and how many exactly?) should be viewed as parts of the mechanism of natural selection. Concerning the MDC account, they note that 'the activities of organisms do not have any particular temporal order', 'any particular rate' or a particular duration. In general, due to the variation that exists within populations they think that 'it is unlikely that natural selection has the degree of organization required by either MDC or Glennan' (p. 338). Prima facie at least, it seems that these points can be raised against Minimal Mechanism too. If natural selection and other population-level mechanisms are not mechanisms in the sense of Minimal Mechanism, then insisting that all mechanistic explanations should conform to Minimal Mechanism would be methodologically misleading.[10]

Another interesting case is the case of developmental mechanisms. In developmental mechanisms the constituents and organisation of the mechanism change in the course of the operation of the mechanism; moreover, developmental mechanisms involve constituents, such as morphogenetic fields, that are diffuse entities, unlike the discrete ones in typical examples of mechanisms given by new mechanists (see McManus 2012). It has yet to be shown that such cases can be captured by Minimal Mechanism or some other inflationary account. In addition, we take it that a lesson of our discussion of constitution in Chapter 8 is that insisting that (some) typical mechanisms in biology are to be understood in terms of Craver's account

---

[10] See Illari and Williamson (2011) for criticisms of Skipper and Millstein's argument. DesAutels (2018) thinks that the mechanism of natural selection can be captured by Minimal Mechanism. Skipper and Millstein view the mechanism of natural selection 'as a chain of temporal steps or stages' (2005, 329) that are causally connected, which is exactly how CM views mechanisms. Interestingly, as Newton's achievement can be seen as introducing a more liberal notion of mechanism, so Charles Darwin introduced a new kind of evolutionary explanation, that is, variational explanation as opposed to transformational explanations of evolution, as, for example, in the case of Lamarck's theory (the distinction is due to Richard Lewontin; see Sober (1984) for discussion). Relatedly, Ernst Mayr has argued that Darwin introduced a new way of thinking into biology, which he called 'population thinking' (1959). As in the case of Newton, Darwin's innovation was met with suspicion by some of his contemporary naturalists and philosophers. Arguably, Darwin's achievement can be interpreted as an introduction of a wholly new type of mechanism to explain evolutionary phenomena. (Darwin never uses the phrase 'mechanism of natural selection'; he usually calls it an 'action', a 'principle', and a 'process' (1859/1964). But such talk – especially talk of the 'process' of selection – can easily be interpreted in terms of CM.)

of constitutive mechanisms puts similar constraints on mechanistic explanation.

The point here is not that there is no way for new mechanists to make the modifications required to capture some of these cases, by clarifying, for example, how processes involving populations and interactions between organisms and environment are to be handled, or what exactly counts as a part. Rather, the point is that a characterisation of mechanism that is not flexible enough to easily accommodate new cases is not useful methodologically. Since we cannot know in advance what mechanisms there are in the world, our general characterisation has to be as open-ended as possible, in order not to constrain scientific practice.

10.4.2.3   Scientific Practice and Inflationary Accounts

Proponents of inflationary accounts might object that talk about entities and activities, for example, cannot function as a constraint on practice in the same way as Old Mechanism constrained seventeenth-century explanations of gravity. To say that mechanisms involve activities, one could argue, should be viewed more as a philosophical gloss on practice, rather than as the elucidation of a concept inherent in practice itself. More generally, new mechanists might object that the purpose of the general characterisation of mechanism is not to guide practice; rather, it is just an abstraction from typical and paradigmatic cases of mechanism and its main purpose is to clarify what a mechanism in general is.

Our answer is as follows: we are mainly interested in identifying a concept-in-use that has a regulatory role within practice. For us, then, the examination of particular cases to derive a general concept is a method to identify the concept-in-use.[11] Since, then, the aim is to understand a concept that functions within practice, the general characterisation cannot just be viewed as a philosophical gloss that is not necessarily relevant for practice, but should identify elements present in practice. And even if we were to agree that talk about activities in no way constrains scientific investigation, the problem of ontological excess content would remain: if talk about activities does not somehow constrain the form of a mechanistic explanation, this would most probably be because it is not really an actual element of practice, but ontological excess content. In view of the sufficiency of CM as an account of practice, this excess content can be omitted from the general characterisation.

---

[11] Note also that cases such as evolutionary and developmental mechanisms are problematic for new mechanists even if they have just the aim of clarifying what a mechanism is.

As we see it, this points to a dissimilarity between current inflationary accounts of mechanism and accounts such as Old Mechanism (or other ontologically inflated views of mechanism). For Boyle and Leibniz, mechanism was a concept of practice but was also viewed in ontological terms. However, in putting constraints on legitimate mechanistic explanations, the ontological content had a role within practice. For new mechanists who adopt inflationary accounts the situation seems different: on the one hand, they do not want to constrain practice; on the other, they give inflationary characterisations of a concept-in-use. But one cannot do both: if practice is not to be constrained, then the ontological content of mechanism should not have an important role within practice; but if it has no such role, it should not be viewed as an element of the concept-in-use.

10.5   Causal Mechanism as a Way Out of the Dilemma

We have seen that new mechanists use a general characterisation of mechanism to flesh out both OT* and MT* and that this leads to a dilemma, because these two theses pull in opposite directions. If we just abandon, or suitably modify, one of these two theses, the dilemma might be resolved. But which one? Since the starting point is to find a general characterisation of mechanism as a concept-in-use, it is clear that MT has priority. After all, one of the primary aims of New Mechanism is to give an account of the role of mechanism in scientific practice. So, the obvious solution is to abandon OT or to weaken it so as to be compatible with a robust version of MT. We can then add another kind of adequacy condition to the general characterisation of mechanism, apart from descriptive and normative adequacy: the general characterisation has to be as minimally ontologically committed as possible. Only such an account would be suitable for making MT as strong as possible. Let us call this *ontological adequacy*.

Causal Mechanism is an account that, as we will now argue, succeeds in being descriptively, normatively and ontologically adequate.

10.5.1   Descriptive Adequacy of CM

Recall that

(CM) A mechanism is a causal pathway, described in theoretical language.

We have already said a lot in the previous chapters about the descriptive adequacy of CM. CM captures a typical use of 'mechanism' in life sciences, which, as we have argued in Chapter 4, identifies a notion of mechanism that is (1) practice-based, (2) common across fields, (3) topic-neutral and (4) diversifiable; it can thus serve as a general characterisation of mechanism in the life sciences.

What CM purports to do is to find the *common denominator* of all uses of the term 'mechanism' in scientific contexts. So, CM is fully compatible with the possibility that in different fields, 'mechanism' can have more specific meanings. Still, CM gives the reason why we can regard all these more specific cases as being instances of a common general concept-in-use. We have then, as we saw in Chapter 3, the following schematic form for more specific kinds of mechanisms:

(P-CM) A mechanism is a causal pathway + X, where X is some external feature of the causal pathway.

The schematic form captures the requirement that it is scientific *practice* itself that will identify what further criteria a causal pathway must fulfil in order to count as a proper mechanism of a particular scientific field. So, an apparent pluralism in scientific practice in how the concept of mechanism is used does not count against CM as a general characterisation that captures a common notion underlying the more specific uses.

10.5.2   Ontological Adequacy of CM

But what exactly *is* a causal pathway? Can we offer CM without fleshing it out in terms, for example, of entities and interactions? But if we do this, we need also to further explain what the entities and interactions are supposed to be, and how they relate to each other. This is where the requirement that the causal pathway should be described in theoretical language comes in. There is simply no more informative and comprehensive way to describe a causal pathway than by reference to the relevant theoretical language. When one of us (S.P.) asked his doctor about the mechanism of Parkinson's, the answer was a description of a pathway. Parkinson's is an incurable progressive neurodegenerative disorder which, because of the depletion of dopamine in the brain, leads to severe motor and movement coordination malfunction and other effects (e.g., dysarthria, bradykinesia and others). As the disease progresses, patients eventually experience severe disability and sometimes dementia in later stages. Well, roughly put, the doctor said, when the supply of dopamine, a key neurotransmitter of motion-related signals, from the substantia nigra (a network of basal ganglia) to the striatum (the part of the brain with neurons that control and coordinate movement) is cut off (for reasons still not entirely clear), there follows a progressive loss of motor functions. Understanding even this rough sketch requires immersion into a theoretical language. It adds nothing by way of understanding to describe this in the language of the metaphysical theory of entities and activities. This sketchy account is a lot more informative than the even sketchier, because abstract and general, account of the new mechanist.

According to CM, the general characterisation is open-ended, in the sense that it avoids any commitment to a specific way to describe a mechanism. The description of the mechanism should in every case be specified in terms of the theoretical language of the appropriate scientific field (or fields). In other words, we should let practice itself decide what are the appropriate theoretical descriptions of a mechanism.

This point can be put as follows: the question 'what is a mechanism?' can be answered, first, by pointing to specific instances of mechanisms in the sciences. If it is asked what all these instances have in common, CM offers a general answer: they are all theoretically describable causal pathways. This is sufficient in order to answer the initial question; any more robust answer would amount to an ontologically inflated characterisation that would unjustifiably constrain scientific practice, a main aim of which is precisely the search for mechanisms.

However, this may still seem unsatisfactory; for, even if by adopting CM we seem not to be committed to any specific account about ontology or the metaphysics of causation, we are still claiming that the pathway is a *causal* one. But then, even if we abstain from saying anything more about what metaphysically grounds causal claims, should we not be committed to a general theory about causation that would distinguish causal pathways from mere sequences of events that are not causally related? This worry can be answered as follows: what is important for the identification of causal pathways in scientific practice is the identification of difference-making relations between the components of the pathway. The identification of the extrinsic pathway of apoptosis, for instance, required the specification of a series of steps, described in molecular terms, that form a causal pathway in the sense that each step makes a difference to what happens next: the initial signal (e.g., the binding of the Fas ligand of T-lymphocytes to the Fas receptor) leads to the activation of the FADD domain of the death receptor, which leads to the recruitment of an adaptor protein, which in turn causes procaspase-8 or 10 to bind to the adaptor protein, which leads to the formation of active caspases 8 and 10, which causes the activation of caspase 3 and the caspase cascade that causes apoptotic cell death.

Let us repeat a key point. While CM is agnostic regarding the meta-physics of causation and abstains from viewing mechanisms in terms of ontological categories, it does not follow that mechanisms qua causal pathways are not parts of the furniture of the world. While a thesis about ontology, this thesis is best viewed as part of a general realist stance concerning science, rather than as part of a new mechanical ontology that science has discovered.

10.5.3   Normative Adequacy of CM

We have identified MT as the main component of New Mechanism and offered CM as a general characterisation of mechanism. Two questions arise now. First, let us grant that explanations in terms of mechanisms are com-monly offered in science. Should we say anything more than this, and in particular should we view the search for mechanisms in (strong) normative terms? Second, even if we accept MT, is this a reason to accept CM? MT and CM (or any general characterisation of mechanism) are logically independent theses: MT says nothing about what a mechanism is, just that science should try to discover them. So, acceptance of MT does not necessitate acceptance of CM. Why, then, not accept both MT and any of the existing general accounts of mechanisms? As should be evident by now, MT is best defended when understood in terms of CM. So, on the one hand, if we already accept MT, CM should be the preferred choice for a characterisation of mechanism. On the other hand, acceptance of CM allows us to understand how MT (even when interpreted in strong normative terms) is plausible in the first place.

The reason that explanations in terms of mechanisms are important is that discovering the causal pathways that in fact produce the phenomena that we investigate is one of the main aims in science. And to do that, what is required is to describe causal pathways in terms of the theoretical language of the particular scientific field (or fields) that studies the phe-nomenon of interest. MT* then becomes:

(MT) Scientists, in investigating the phenomena, should search for mechanisms responsible for them, where a mechanism is *a causal pathway described in theoretical language*.

10.6   The Triviality Problem

How informative, really, is MT understood in terms of CM? MT may seem trivial and even vacuous due precisely to the refusal of methodolog-ical mechanists to commit themselves to some robust version of OT. The underlying assumption behind this triviality problem, as we will call it, is that in order to have normative force, MT should exclude alternative methodological viewpoints. But if it remains ontologically non-committal, it seems that MT is too minimal to do this. The problem, then, is that by adopting CM as a characterisation of mechanism as concept-in-use, we in effect trivialise MT.

However, MT is far from a trivial thesis. It has both a negative and a positive role in guiding scientific practice. Historically, at least, there certainly have been alternative methodological viewpoints to mechanism. A central motivation for the introduction of the corpuscularian philosophy in the seventeenth century was precisely the presence of such an alternative methodology, namely, the explanation of phenomena in terms of substantial forms, Aristotelian powers and other unintelligible – from the point of view of old mechanists – metaphysical entities.[12] In early twentieth-century biology, vitalism also constituted an alternative non-mechanistic methodological viewpoint (see Allen 2005). Even if seventeenth-century Aristotelianism and early twentieth-century vitalism are usually viewed as ontological theses, they are also theses about biological methodology and the forms that biological explanation should take.

What is common among non-mechanistic viewpoints is their recourse to what seem for the point of view of mechanists as unintelligible notions (e.g., sui generis powers, substantial forms and entelechies) that incorporate some kind of teleology.[13] If, for example, an Aristotelian explains how X produces Y by saying that X has the power to produce Y without saying anything about the way (i.e., the causal pathway) that Y is in fact produced, or if a vitalist explains the course of development by postulating an entelechy, then, mechanists think, no explanation has really been given. What is missing is the means (i.e., the causal pathway) by which these powers and entelechies do their causal work.[14]

Note also that for MT to be informative it is not important that there actually exists a rival scientific tradition, as was the case in the seventeenth

---

[12] See our discussion of Boyle in Chapter 1.
[13] As Brandon (1984) stresses, there is a kind of teleology that is compatible with mechanism, namely, the teleology involved in evolutionary explanations of adaptations, where an adaptation is explained in terms of its effects on the organism that possesses it. Since such explanations involve the mechanism of natural selection, the teleology in question is not opposed to mechanistic methodology.
[14] Except, of course, in cases like fundamental physics, where there can be no mechanism that mediates between cause and effect. Note also that commitment to this explanatory norm is not to deny the existence of powers; powers may be a way to ground causation in a mechanism, but what explains a phenomenon is the mechanism itself.

century. Even in the absence of this, in regulating scientific practice, part of the role of MT is to block certain kinds of explanations among biologists or to guard against implicit assumptions. In the following quotation by the cell biologist Richard Lockshin we see this negative role of MT in action:

> a cell . . . *neither plans its future nor considers its relationship to the organism.* In the *mechanistic view of cell biology*, biochemical and biophysical changes within the cytoplasm beget adjustments that activate autophagy, apoptosis, or other responses. (Lockshin 2016, 14; emphasis added)

We take the point that Lockshin makes here to be that the mechanistic viewpoint is a non-teleological viewpoint, in the sense that it does not allow explaining biological phenomena by attributing folk psychological properties to biological entities such as cells. Instead, what must be explained is how changes in the cell give rise to subsequent events; and to do this, the causal pathways that produce the phenomena must be identified. So, even if minimal as a methodological norm, the consequences of MT for scientific practice can be quite drastic.

Apart from its negative role in blocking certain kinds of explanation, there is also a positive role that MT has. Mechanisms offered in science are often incomplete, in the sense that various details of the causal pathway can be unknown. For example, when Kerr et al. (1972) first proposed the existence of apoptosis as a mechanism of cell death, the biochemical mechanisms responsible for the apoptotic process were completely unknown. In Chapter 3 (Section 3.5.2) we mentioned a vertical and a horizontal dimension that are involved in making theoretical descriptions more detailed. Take again the cytological description of apoptosis. The horizontal dimension involves identifying more details at the cytological level of organisation, whereas the vertical dimension involves identifying processes at lower levels of organisation. In instructing scientists to always identify causal pathways, MT directs them to fill in the missing details in incomplete descriptions of mechanisms.[15] To the extent that existing descriptions of mechanisms qua causal pathways can always be made more

---

[15] As we saw in the case of apoptosis, incomplete knowledge of the details of a causal pathway does not prevent scientists from identifying it and describing it as a new 'mechanism'. The importance of filling in the missing details in incomplete descriptions of mechanisms is a point that new mechanists have emphasised; see, for example, Darden's (2002) distinction between mechanism schemas and mechanism sketches (on schemas and sketches, see also Machamer et al. (2000); on various strategies for discovering mechanisms, see Craver and Darden (2013)).

detailed, this positive role of MT is pervasive in science. Note also that to take scientific practice seriously is not only to account for concepts such as mechanism that are central in practice but also to account for central methodological norms such as MT. By showing that MT is a compelling view, CM allows us to appreciate why searching for mechanisms is widespread in science in a way that ontologically inflated accounts cannot.

Pervasiveness should not be confused with triviality. The fact that the mechanistic methodological viewpoint, in the sense we have been explicating it, is so widespread as to perhaps seem trivial cannot lead to a criticism of MT as a thesis that adequately describes what scientists are doing: the descriptive adequacy of a philosophical account of a piece of scientific methodology should certainly count in favour of the philosophical account in question. Last, if the accusation of triviality concerns the ontologically minimal account of mechanism that MT incorporates, then let us note once again that mechanism in science cannot mean anything more robust than CM if it is going to be a concept useful in practice: far from being a trivial thesis, MT advocates searching for mechanisms irrespective of what the fundamental ontological structure of the world really is.[16]

In sum: if we accept, together with new mechanists, that a common notion of mechanism is present across scientific fields, then CM is the best candidate for a general characterisation of this notion. CM captures the unifying character of the concept of mechanism; it is an account that remains close to practice, without incorporating elements that do not seem to be needed by practicing scientists (such as a commitment to powers or activities); at the same time, it is diversifiable, as it can easily be adapted to account for how the concept functions in specific scientific fields; in addition, in showing why MT is central in science, CM succeeds in being

---

[16] A possible worry here about CM is that it (and a fortiori MT) seems almost vacuous, since everything is (or has) a mechanism: Are there things-in-the-world that are not mechanisms? The worry, then, is that 'mechanism' becomes a concept devoid of real empirical content. Note, by way of reply, that this kind of worry can be effective, if at all, against 'thicker' accounts of mechanism too. It is not clear, for instance, what does not count as a mechanism on Glennan's Minimal Mechanism account – though on Glennan's earlier views there are non-mechanisms (only) at the level of fundamental physics. Be that as it may, our answer would simply be: something is not a mechanism in the CM sense if it is not a causal pathway. More importantly, however, the objection has a bite against MT only if MT is taken to be a metaphysical thesis, which it is not. As such, the proper contrast (as noted above) is not what-in-the-world-is-not-a-mechanism versus what-in-the-world-is-a-mechanism, but rather: Are there alternative methodological standpoints that explain non-mechanistically?

non-trivial and thus illuminating and informative both as a general account of practice as well as a concept central to scientific practice itself. Acceptance of CM resolves the dilemma faced by New Mechanism, since to accept CM means to abandon a robust version of the ontological tenet. New Mechanism as a framework to think about science is then best viewed as built around a primarily methodological thesis; New Mechanism is Methodological Mechanism.

Finale

The central question of this book has been: How should we characterise the concept of mechanism as a concept-in-use of scientific/biological practice? We have argued that the most appropriate general characterisation is what we have called Causal Mechanism. In this finale we examine to what extent CM is a properly 'mechanistic' thesis, that is, to what extent it can be seen as a descendant of the original notion of mechanism developed in seventeenth century. In so doing, we examine possible extensions of the seventeenth-century notion of mechanism and discuss whether they can be used to characterise mechanism as a concept-in-use.

Our strategy will be to examine how we can extend the original notion of mechanism by abstracting from the details of physical theory, while at the same time retaining enough content so that the resulting notions can be seen as more general notions of the same family of concepts. Following this strategy, we identify two central and independent conditions that a biological explanation has to satisfy in order to count as mechanistic, both of which were central in Old Mechanism: the condition of intelligibility (which says that a mechanistic explanation should not resort to unintelligible principles such as vital powers) and the condition of the priority of the parts over the whole. In order to clarify this second condition, we develop the notion of mechanistic reduction, which is motivated by the historical discussion of Chapter 2. We argue that mechanistic reduction entails causal modularity and thus that the failure of causal modularity is an indication that the parts have to be seen as in some sense dependent on the whole.

We use our two conditions to distinguish between two notions of mechanism: a more narrow one that incorporates both the intelligibility condition and the condition of the priority of the parts and a broader one that incorporates only the intelligibility condition and is thus a weakened form of mechanism. We claim that an account of mechanism as a concept-in-use requires the weakened notion, which when viewed in terms of CM has nevertheless enough content so that it can be seen as a descendant of the original concept.

Two Conditions for Mechanistic Explanation

As we have seen in Chapter 1, a central consideration in the context of Old Mechanism is what we can call the intelligibility argument. Here is how Leibniz puts it in his essay 'Against Barbaric Physics':

> That physics which explains everything in the nature of body through number, measure, weight, or size, shape and motion, which teaches that nothing is moved naturally except through contact and motion, and so teaches that, in physics, everything happens mechanically, that is, intelligibly, this physics seems excessively clear and easy. (1989, 312)

Here, Leibniz equates the mechanical with the intelligible. Indeed this argument, that is, that only an explanation in terms of the mechanical affections of parts of matter is intelligible, whereas an explanation in terms of scholastic powers, substantial forms and real qualities is obscure ('barbaric' according to Leibniz), is central in many thinkers of the seventeenth century. Descartes and Boyle, for example, both argue that only a mechanical explanation is intelligible. In Old Mechanism, then, explanations of phenomena in terms of sui generis powers of things are non-mechanical.

The condition of intelligibility is not just a negative thesis; that is, it does not just say what a mechanistic explanation should not do. It has some positive content too. This positive content can be seen very clearly in Boyle's writings examined in Chapter 1; a mechanistic explanation, for Boyle, has to state how exactly the cause operates to bring about the effect. In other words, in giving a mechanistic explanation we have to specify the causal steps (or the causal pathway) leading from an initial cause to an effect.

We can retain this condition on mechanistic explanation, which we will call the *condition of intelligibility*, even if we abandon the specific details of seventeenth-century mechanistic physics. So, we can expand the concept of mechanism to allow for chemical and other kinds of interactions among the components of a pathway. We thereby satisfy the spirit of Old Mechanism as long as in doing so we do not reintroduce the sui generis powers of scholastic Aristotelianism.[1]

---

[1] What is important here is that biological phenomena should not be explained in terms of illegitimate forms of teleology, which was a central aspect of scholastic powers; see our discussion in Chapter 10. The use of teleology and sui generis vital forces to contrast mechanical with non-mechanical

As we saw in Chapter 2, mechanistic explanation has historically been associated with a second idea, that is, that the properties and behaviours of the parts are in some sense independent of the whole, and so the properties and behaviour of the whole can be explained in terms of its parts. We have here a second condition for mechanistic explanation, that is, the priority of the parts over the whole. This form of explanation was central in Old Mechanism: for thinkers such as Boyle and Descartes, the capacity of a clock to tell time and the capacity of fire to burn wood are both explained in terms of the motions of the parts of matter that make up the clock or fire. This condition too need not be viewed just in the context of seventeenth-century physics.

The rejection of these conditions leads to non-mechanistic explanations. The rejection of the first condition, that is, the positing of sui generis vital powers and final causality, leads to vitalism. Such a position, for example, was adopted by Hans Driesch at the end of the nineteenth century, who posited what he called 'entelechies' that somehow regulate the operation and development of organisms. The rejection of the priority of the parts over the whole leads to holism (or 'organicism'). Twentieth-century holists argued that biological systems, despite being, in contrast to the vitalistic doctrine, 'mere' physicochemical systems, form a unified whole and thus are not 'mechanisms'. According to holists, higher biological levels of organisation are indispensable in giving biological explanations of phenomena such as reproduction, purposeful reaction to stimuli and complex self-regulation. As Allen puts it, '[i]t is in their appreciation of the concept that each level of organization in a complex system has its own special properties, and that these must by studied by techniques appropriate for that level, that holists differ in one significant way from Mechanists' (2005, 168).

We have thus identified two conditions that a biological explanation has to satisfy in order to count as mechanistic. These two conditions are logically independent. For example, holists adopted the first condition but not the second. Although vitalism as developed by Driesch rejects both conditions, it is also possible to think that the parts have priority over the whole, which can be explained in terms of them, but to accept that the parts have vital powers.[2]

---

explanation is justified historically; see our discussion of Kant's views and Broad's Substantial Vitalism in Chapter 2. Note also that current versions of neo-Aristotelianism are not at odds with the intelligibility condition, since they do not postulate sui generis vital powers.

[2] Bertoloni Meli gives the example of the eighteenth-century anatomist Xavier Bichat, who explained the activity of organs in terms of their component parts, but took it that those parts had vital

Given the above, we can ask: Do contemporary notions of mechanism satisfy these conditions? The first condition seems the most easy to satisfy: no biologist nowadays thinks that we need to introduce sui generis vital powers to understand the behaviour of biological systems. When we turn to the second condition, however, things are more difficult. Can a whole always be explained in terms of the properties of parts that are taken not to depend in some sense on the whole, or are there cases where the whole has some degree of autonomy with respect to the parts?

Causal Modularity

We can answer this question by considering the notion of causal modularity. According to some philosophers, for an explanation to be mechanistic, or for a system to be regarded as a mechanism, it should have a modular structure: it should be in principle possible to change a particular causal relationship within the system (e.g., by removing one of the components of the system) without changing other causal relationships in the system (see Woodward 2002; Menzies 2012). This kind of modularity as 'independent disruptability' (Hausman & Woodward 1999) makes possible the decomposition of the whole effect of a causal system into the independent causal contributions of its constituents. Modularity is thought to capture the sense in which the behaviour of the components of the mechanism are independent of the behaviour of the mechanism as a whole. The idea is that only if this independence obtains can we explain the whole in terms of the parts. Modularity has thus been viewed as a necessary condition for mechanistic explanation.

Woodward, in particular, takes mechanistic explanation to be a species of causal explanation. He formulates modularity within his interventionist framework of causation, which we discussed in Chapter 7. A mechanism is

properties not found in non-living matter. Bertoloni Meli uses the example of Bichat to criticise Bechtel's claim that 'Bichat was pursuing a program of mechanistic explanation' (Bechtel 2006, 45). He stresses that in fact Bichat 'actively opposed the mechanistic program because he deemed it erroneous'. He notes that 'if the defining feature of a mechanism is that it operates "in virtue of its component parts" [as Bechtel argues], Bechtel should argue that Aristotle and Galen too, despite their teleology, in crucial respects were "pursuing a program of mechanistic explanation" because they "attempted to explicate the behavior" of bodies in terms of the organs "out of which they were constructed"' (Bertoloni Meli 2019, 6). Although Bertoloni Meli makes this point in the context of arguing that an approach such as Bechtel's 'may be adequate for systematic concerns and analyses of the role of mechanism in biology, more sophisticated tools are needed for a meaningful historical analysis' (p. 6), we take the point to be that an adequate characterisation of mechanism and mechanistic explanation has to incorporate both the condition of intelligibility and the condition of the priority of parts.

a set of components, which are characterised by variables that stand in causal dependence relations to each other. Woodward takes a representation of these variables standing in causal relations to be modular 'to the extent that each of the individual G_i [generalisations that describe the causal relationships among the components of the system] remain at least somewhat stable under interventions that change the other G_i' (Woodward 2013, 51). Woodward thinks that the requirement that causal relationships within the mechanism have to be modular captures the sense in which a mechanistic explanation explains the behaviour of the whole in terms of the intrinsic behaviour of the parts. He cites biologists George von Dassow and Ed Munro, who claim that:

> Mechanism, per se, is an explanatory mode in which we describe what are the parts, how they behave intrinsically, and how those intrinsic behaviors of parts are coupled to each other to produce the behavior of the whole. This common sense definition of mechanism implies an inherently hierarchical decomposition; having identified a part with its own intrinsic behavior, that part may in turn be treated as a whole to be explained. (von Dassow & Munro 1999, 309)

According to Woodward, in a case of a modular system:

> each such subset of causally related components continues to be governed by the same set of causal relationships, independently of what may be happening with components outside of that subset, so that the behaviour is (in this respect) 'intrinsic' to that subset of components . . . modularity seems to me to capture at least part of what is involved in their notion of 'intrinsicness'. (2013, 51)

Note here exactly what modularity requires. It does not require that, were we to intervene to disrupt a causal relationship between a pair of variables, the mechanism would produce the same result as before the intervention; in the case of such a disruption, the production of the outcome of the mechanism would be disrupted too. The claim, rather, is that the disruption of a causal dependence, for example, by removing a component, would not disrupt what other causal dependencies there are in the system.

The important point for our discussion is that mechanistic reduction implies modularity. That is, if modularity captures the sense of 'intrinsicness', as Woodward claims, which is the central feature of mechanistic reduction as explained above, then, if we have a system such as a mechanical clock where organisation just concerns the spatial relations among independently existing components with intrinsic properties, such a system is modular. But then, if a system fails to be modular, this means that mechanistic reduction fails too.

Many biological systems do not satisfy the condition of modularity as independent disruptability (see Mitchell 2008).[3] In such 'robust' systems, a change in some causal interactions within the system may lead to a restructuring of the system. So, failure of modularity shows that the nature of biological components is not entirely independent of the whole in which they are located. Rather, some of their properties are dependent on their being components of a certain whole. As Woodward puts it, in non-modular systems 'how some of the components behave depends in a global, "extrinsic" way on what is going on in other components' (2013, 54).

Moreover, parts of biological systems can depend on the whole for their existence. For example, the cell regulates various parameters that are necessary for the operation of cellular mechanisms. Thus, the existence of these mechanisms depends upon the existence of a properly functioning cell. In general, organisation in biological systems does not just concern a network of spatial relations, as in the case of a mechanical clock. In the case of a cell, organisation is not imposed on the mechanism from 'outside', so to speak, but is itself in part the result of the operation of the mechanism.[4]

Causal Mechanism Is Mechanism Enough

Some of the notions we have examined in Chapters 1 and 2 satisfy both conditions of intelligibility and priority of parts, that is, Cartesian mechanism, mechanical mechanism (i.e., the more liberal post-Newtonian notion), Ewing's quasi-mechanical mechanism and Broad's Pure Mechanism and Biological Mechanism. Quasi-mechanical and Biological Mechanism are here the more general notions, that is, the ones that do not make any specific assumptions about the underlying physics or chemistry. But can such concepts capture the mechanism talk that is ubiquitous in life sciences?

---

[3] More generally, according to Woodward, 'many biological systems require explanations that are relatively non-mechanical or depart from expectations one associates with the behaviour of machines' (2013, 39), for example, dynamical systems explanations.

[4] Biological systems such as cells and organisms are what have been called 'autonomous' systems. An autonomous system is 'a far-from-equilibrium system that constitutes and maintains itself establishing an organisational identity of its own, a functionally integrated (homeostatic and active) unit based on a set of endergonic-exergonic couplings between internal self-constructing processes, as well as with other processes of interaction with its environment' (Ruiz-Mirazo et al. 2004).

Mechanism as concept-in-use seems to be a very liberal notion, applied to all sorts of systems irrespective of their causal organisation. For example, Lakhani et al. (2009, 4) write about the 'strongly biomedical concept of disease' they adopt, that it is 'a mechanistic model that regards the body as a machine with repairable or replaceable parts. It looks for specific underlying biological causes and places a high emphasis on the scientific evidence-base for untangling cause and effect in both the disease and its treatment, because this is important for patient care and prognosis' (p. 4). But although they adopt a mechanistic model, 'it is a complex model with multiple parts that interconnect. A change in one area is likely to affect another. Thus maintaining homeostasis is not a simple single feedback loop and it is perfectly acceptable that a new equilibrium is achieved under a new set of circumstances, a new baseline; you do not have to return to the original state' (p. 4). On this account, to adopt a 'mechanistic model' concerning the body is just to say that the body contains various interacting parts that underlie bodily functions and sometimes result in diseases, such that it is possible to identify cause and effect relationships and to intervene to treat parts that malfunction.[5]

A way to capture the more liberal notion of mechanism is to reject the condition of the priority of the parts. So we are left only with the intelligibility condition. Would that be enough for mechanism? One might be tempted to add the claim that though modularity (or priority of the parts) doesn't hold, something near it does, that an activity of a whole is explained in terms of the organised entities and activities that constitute it. However, this leads to a form of mechanism where a whole is explained in terms of its parts, but the parts may not be independent of the whole. The main problem here is that we lose the clear contrast with what a non-mechanistic explanation would be. Broad's Substantial Vitalism does certainly count as a non-mechanistic position on this view, but what about Emergent Vitalism, or the holist and organicist positions of the early twentieth century? If these are to be counted as mechanistic too, why use the term 'mechanism' at all? There is here the danger that the concept of mechanism may become vacuous. In addition, it becomes difficult to view this weakened notion as a descendant of the original notion of mechanism.

Our alternative is to view mechanism as combining intelligibility with Causal Mechanism. As we view mechanism as a concept-in-use of

---

[5] For a detailed examination and criticism of the claim that the cell can be seen as a machine, see Nicholson (2019).

biological practice, the aim of this practice is not to explain a whole in terms of its components but to trace the causal steps, that is, to identify the causal pathway, leading from an initial cause to an effect.[6] The centrality of the concept in practice is an indication that the notion is not vacuous, but has an important methodological role (as we have argued in Chapter 10). Adopting CM fully embraces the processual character of biological mechanisms; moreover, it does not blur the distinction between mechanism and holist or emergentist positions; last, it can easily be viewed as a descendant of the original seventeenth-century concept of mechanism. Causal Mechanism is Mechanism Enough!

---

[6] Nicholson's (2011) analysis of the concept of mechanism in biology shares some similarities with our approach. Nicholson distinguishes between what he calls 'machine mechanism', which concerns 'the internal workings of a machine-like structure', and 'causal mechanism', which concerns 'the causal explanation of a particular phenomenon'. He claims that new mechanists like Craver, Darden and Bechtel conflate these two senses of mechanism and 'inappropriately endow causal mechanisms with the ontic status of machine mechanisms, and this invariably results in problematic accounts of the role played by mechanism-talk in scientific practice' (p. 152).

References

Alberts, B., Johnson, A., Lewis, J., Raff, M., Roberts, K. & Walters, P. (2014). *Molecular Biology of the Cell*, 6th ed. New York: Garland Science.

Allen, G. E. (2005). Mechanism, vitalism and organicism in late nineteenth and twentieth-century biology: the importance of historical context. *Studies in History and Philosophy of Biological and Biomedical Sciences*, 36, 261–83.

Andersen, H. (2012). The case for regularity in mechanistic causal explanation. *Synthese*, 189, 415–32.

(2014a). A field guide to mechanisms: part I. *Philosophy Compass*, 9, 274–83.

(2014b). A field guide to mechanisms: part II. *Philosophy Compass*, 9, 284–93.

Armstrong, D. M. (1983). *What Is a Law of Nature?* Cambridge: Cambridge University Press.

(1997). Singular causation and laws of nature. In J. Earman and J. Norton, eds., *The Cosmos of Science*. Pittsburgh, PA: University of Pittsburgh Press, pp. 498–511.

Baetu, T. M. (2019). *Mechanisms in Molecular Biology*, Cambridge Elements. Cambridge: Cambridge University Press.

Baron, R. M. & Kenny, D. A. (1986). The moderator-mediator variable distinction in social psychological research: conceptual, strategic, and statistical considerations. *Journal of Personality and Social Psychology*, 51, 1173–82.

Bartholomew, M. (2002). James Lind's treatise of the scurvy (1753). *Postgraduate Medical Journal*, 78, 695–6.

Baumgartner, M. & Casini, L. (2017). An abductive theory of constitution. *Philosophy of Science*, 84, 214–33.

Baumgartner, M. & Gebharter, A. (2016). Constitutive relevance, mutual manipulability, and fat-handedness. *The British Journal for the Philosophy of Science*, 67, 731–56.

Bechtel, W. (2006). *Discovering Cell Mechanisms: The Creation of Modern Cell Biology*. Cambridge: Cambridge University Press.

(2008). *Mental Mechanisms: Philosophical Perspectives on Cognitive Neuroscience*. New York: Routledge.

(2017). Explicating top-down causation using networks and dynamics. *Philosophy of Science*, 84, 253–74.

Bechtel, W. & Abrahamsen, A. (2005). Explanation: a mechanistic alternative. *Studies in History and Philosophy of Biological and Biomedical Sciences*, 36, 421–41.

Bechtel, W. & Richardson, R. C. (2010) [1993]. *Discovering Complexity: Decomposition and Localization as Strategies in Scientific Research*, 2nd ed. Cambridge, MA: MIT Press/Bradford Books.

Beiser, F. (2005). *Hegel*. New York: Routledge.

Bennett, J. (2003). *A Philosophical Guide to Conditionals*. Oxford: Oxford University Press.

Bertoloni Meli, D. (2019). *Mechanism: A Visual, Lexical, and Conceptual History*. Pittsburgh, PA: University of Pittsburgh Press.

Bird, A. (2007). *Nature's Metaphysics: Laws and Properties*. Oxford: Oxford University Press.

Boas, M. (1952). The establishment of the mechanical philosophy. *Osiris*, 10, 412–541.

Bogen, J. (2005). Regularities and causality; generalizations and causal explanations. *Studies in History and Philosophy of Biological and Biomedical Sciences*, 36, 397–420.

Boniolo, G. & Campaner, R. (2018). Molecular pathways and the contextual explanation of molecular functions. *Biology and Philosophy*, 33, 24. https://doi.org/10.1007/s10539-018-9634-2.

Bork, A. (1967). Maxwell and the electromagnetic wave equation. *American Journal of Physics*, 35, 83–9.

Boyle, R. (1991). *Selected Philosophical Papers of Robert Boyle*, ed. M. A. Stewart. Indianapolis, IN: Hackett Publishing Company.

Brandon, R. N. (1984). Grene on mechanism and reductionism: more than just a side issue. In P. Asquith and P. Kitcher, eds., *PSA: Proceedings of the Biennial Meeting of the Philosophy of Science Association*, vol. 2. East Lansing, MI: Philosophy of Science Association, pp. 345–53.

   (1990). *Adaptation and Environment*. Princeton, NJ: Princeton University Press.

Breitenbach, A. (2006). Mechanical explanation of nature and its limits in Kant's *Critique of Judgement*. *Studies in History and Philosophy of Biological and Biomedical Sciences*, 37, 694–711.

Broad, C. D. (1925). *Mind and Its Place in Nature*. London: Routledge and Kegan Paul.

Brown, S. R. (2003). *Scurvy: How a Surgeon, a Mariner, and a Gentleman Solved the Greatest Medical Mystery of the Age of Sail*. Chichester: Summersdale Publishers.

Bugianesi, E., McCullough, A. & Marchesini, G. (2005). Insulin resistance: a metabolic pathway to chronic liver disease. *Hepatology*, 42, 987–1000.

Cairrão F. & Domingos, P. M. (2010). Apoptosis: molecular mechanisms. In *Encyclopedia of Life Sciences*. Chichester: John Wiley & Sons.

Campaner, R. (2006). Mechanisms and counterfactuals: a different glimpse of the (secret?) connexion. *Philosophica*, 77, 15–44.

Carpenter, K. J. (2012). The discovery of vitamin C. *Annals of Nutrition and Metabolism*, 61, 259–64.

Cartwright, N. D. (1989). *Nature's Capacities and Their Measurement*. Oxford: Clarendon Press.

Casini, L. (2016). Can interventions rescue Glennan's mechanistic account of causality? *British Journal for the Philosophy of Science*, 67, 1155–83.

Chisholm, R. (1946). The contrary-to-fact conditionals. *Mind*, 55, 289–307.

Clarke, B., Gillies, D., Illari, P., Russo, F. & Williamson, J. (2014). Mechanisms and the evidence hierarchy. *Topoi*, 33, 339–60.

Clarke, P. G. H. & Clarke, S. (1996). Nineteenth century research on naturally occurring cell death and related phenomena. *Anatomy and Embryology*, 193, 81–99.

Couch, M. B. (2011). Mechanisms and constitutive relevance. *Synthese*, 183, 375–88.

Cox, D. R. (1986). Comment. *Journal of the American Statistical Association*, 81, 963–4.

(1992). Causality: some statistical aspects. *Journal of the Royal Statistical Society Series A*, 155, 291–301.

Cox, D. R & Wermuth, N. (2001). Some statistical aspects of causality. *European Sociological Review*, 17, 65–74.

Craver, C. F. (2001). Role functions, mechanisms and hierarchy. *Philosophy of Science*, 68, 31–55.

(2007a). *Explaining the Brain: Mechanisms and the Mosaic Unity of Neuroscience*. Oxford: Oxford University Press.

(2007b). Constitutive explanatory relevance. *Journal of Philosophical Research*, 32, 3–20.

(2013). Functions and mechanisms: a perspectivalist view. In P. Huneman, ed., *Functions: Selection and Mechanisms*. Dordrecht: Springer, pp. 133–58.

Craver, C. F. & Bechtel, W. (2007). Top-down causation without top-down causes. *Biology and Philosophy*, 22, 547–63.

Craver, C. F. & Darden, L. (2005). Introduction. *Studies in History and Philosophy of Biological and Biomedical Sciences*, 36, 233–44.

(2013). *In Search of Mechanisms: Discoveries across the Life Sciences*. Chicago: University of Chicago Press.

Craver C. & Tabery, J. (2015). Mechanisms in science. In E. N. Zalta, ed., *The Stanford Encyclopedia of Philosophy* (Summer 2019 edition), https://plato.stanford.edu/archives/sum2019/entries/science-mechanisms/.

Cummins, R. (1975). Functional analysis. *Journal of Philosophy*, 72, 741–64.

Darden, L. (2002). Strategies for discovering mechanisms: schema instantiation, modular subassembly, forward/backward chaining. *Philosophy of Science*, 69, S354–S365.

(2006). *Reasoning in Biological Discoveries*. Cambridge: Cambridge University Press.

Darwin, C. (1859/1964). *Origin of Species*. Cambridge, MA: Harvard University Press.

Dawid, P. (2000). Causal inference without counterfactuals. *Journal of the American Statistical Association*, 95, 407–24.

DesAutels, L. (2011). Against regular and irregular characterizations of mechanisms. *Philosophy of Science*, 78, 914–25.

(2018). Mechanisms in evolutionary biology. In S. Glennan and P. Illari, eds., *The Routledge Handbook of Mechanisms and Mechanical Philosophy*. New York: Routledge, pp. 296–307.

Descartes, R. (1982). *Principles of Philosophy*, trans. V. R. Miller and R. P. Miller. Dordrecht: D. Reidel Publishing Company.

(2004). *René Descartes: The World and Other Writings*, ed. S. Gaukroger. Cambridge: Cambridge University Press.

Dowe, P. (2000). *Physical Causation*. Cambridge: Cambridge University Press.

Dretske, F. I. (1977). Laws of nature. *Philosophy of Science*, 44, 248–68.

Ellis, B. (2001). *Scientific Essentialism*. Cambridge: Cambridge University Press.

Ellis, H. M. & Horvitz, H. R. (1986). Genetic control of programmed cell death in the nematode C. elegans. *Cell*, 44, 817–29.

Eronen, M. I. (2015). Levels of organization: a deflationary account. *Biology and Philosophy*, 30, 39–58.

Eronen, M. I. & Brooks, D. S. (2018). Levels of organization in biology. In E. N. Zalta, ed., *The Stanford Encyclopedia of Philosophy* (Spring 2018 edition), https://plato.stanford.edu/archives/spr2018/entries/levels-org-biology/.

Ewing, A. C. (1969). *Kant's Treatment of Causality*. Hamden, CT: Archon Books.

Fink, S. L. & Cookson, B. T. (2005). Apoptosis, pyroptosis, and necrosis: mechanistic description of dead and dying eukaryotic cells. *Infection and Immunity*, 73, 1907–16.

Franklin-Hall, L. (2016). New mechanistic explanation and the need for explanatory constraints. In K. Aizawa and C. Gillett, eds., *Scientific Composition and Metaphysical Ground*. London: Palgrave Macmillan, pp. 41–74.

Funk, C. (1912). The etiology of the deficiency diseases. *The Journal of State Medicine*, 20, 341–68.

Gardner, D. G. & Shoback, D. (2017). *Greenspan's Basic & Clinical Endocrinology*, 10th ed. New York: McGraw-Hill.

Garson, J. (2013). The functional sense of mechanism. *Philosophy of Science*, 80, 317–33.

(2018). Mechanisms, phenomena, and functions. In S. Glennan and P. Illari, eds., *The Routledge Handbook of Mechanisms and Mechanical Philosophy*. New York: Routledge, pp. 104–15.

Gilbert, S. F. (2010). *Developmental Biology*, 9th ed. Sunderland, MA: Sinauer Associates.

Gillett, C. (2006). The metaphysics of mechanisms and the challenge of the new reductionism. In M. Schouten and H. L. de Jong, eds., *The Matter of the Mind*. Oxford: Blackwell, pp. 76–100.

(2013). Constitution, and multiple constitution, in the sciences: using the neuron to construct a starting framework. *Minds and Machines*, 23, 309–37.

(2016). *Reduction and Emergence in Science and Philosophy*. Cambridge: Cambridge University Press.

Gillies, D. (2011). The Russo-Williamson thesis and the question of whether smoking causes heart disease. In P. Illari, F. Russo and J. Williamson, eds., *Causality in the Sciences*. Oxford: Oxford University Press, pp. 110–25.

(2017). Mechanisms in medicine. *Axiomathes*, 27, 621–34.

(2019). *Causality, Probability and Medicine*. London: Routledge.

Ginsborg, H. (2004). Two kinds of mechanical inexplicability in Kant and Aristotle. *Journal of the History of Philosophy*, 42, 33–65.

Glennan, S. (1992). Mechanisms, models and causation. PhD dissertation, University of Chicago.

(1996). Mechanisms and the nature of causation. *Erkenntnis*, 44, 49–71.

(2002). Rethinking mechanistic explanation. *Philosophy of Science*, 69, S342–S353.

(2010). Ephemeral mechanisms and historical explanation. *Erkenntnis*, 72, 251–66.

(2011). Singular and general causal relations: a mechanist perspective. In P. Illari, F. Russo and J. Williamson, eds., *Causality in the Sciences*. Oxford: Oxford University Press, pp. 789–817.

(2017). *The New Mechanical Philosophy*. Oxford: Oxford University Press.

(2021). Corporeal composition. *Synthese*, 198, 11439–62.

Glennan, S. & Illari, P., eds. (2018a). *The Routledge Handbook of Mechanisms and Mechanical Philosophy*. New York: Routledge.

(2018b). Introduction: mechanisms and mechanical philosophies. In S. Glennan and P. Illari, eds., *The Routledge Handbook of Mechanisms and Mechanical Philosophy*. New York: Routledge, pp. 1–9.

Goodman, N. (1947). The problem of counterfactual conditionals. *Journal of Philosophy*, 44, 113–28.

Grene, M. (1971). Reducibility: another side issue? In M. Grene, ed., *Interpretations of Life and Mind*. New York: Humanities Press, pp. 14–37 (reprinted in R. S. Cohen and M. W. Wartofsky, eds., The Understanding of Nature, Boston Studies in the Philosophy of Science, vol. 23 [Dordrecht: Reidel, 1974], pp. 53–73).

Hall, N. (2004). Two concepts of causation. In J. Collins, N. Hall and L. Paul, eds., *Causation and Counterfactuals*. Cambridge, MA: MIT Press, pp. 225–76.

Harbecke, J. (2010). Mechanistic constitution in neurobiological explanations. *International Studies in the Philosophy of Science*, 24, 267–85.

Harinen, T. (2018). Mutual manipulability and causal inbetweenness. *Synthese*, 195, 35–54.

Harré, R. (1970). *The Principles of Scientific Thinking*. London: Macmillan.

(1972). *The Philosophies of Science: An Introductory Survey*. Oxford: Oxford University Press.

(2001). Active powers and powerful actors. In A. O'Hear, ed., *Philosophy at the New Millennium*. Cambridge: Cambridge University Press, pp. 91–109.

Hausman, D. M. & Woodward, J. (1999). Independence, invariance and the Causal Markov condition. *British Journal for the Philosophy of Science*, 50, 521–83.

Hegel, G. W. F. (1832/1991). *The Encyclopaedia Logic, Part I of the Encyclopaedia of Philosophical Sciences with the Zusätze*, trans. T. F. Geraets, W. A. Suchting and H. S. Harris. Indianapolis: Hackett Publishing Company.

(2002). *Science of Logic*, trans. A. V. Miller. London: Routledge.

Hertz, H. (1894/1955)]. *The Principles of Mechanics Presented in a New Form*, first English trans. 1899, reprinted by Dover. New York: Dover Publications.

Hill, B. (1965). The environment of disease: association or causation? *Proceedings of the Royal Society of Medicine*, 58, 295–300.

Hogben, L. (1930). *The Nature of Living Matter*. London: Kegan Paul, Tranch, Trubner and Co.

Holland, P. (1986). Statistics and causal inference. *Journal of the American Statistical Association*, 81, 945–60.

(1988). Comment: causal mechanism or causal effect: which is best for statistical science? *Statistical Science*, 3, 186–8.

Horwich, P. (1987). *Asymmetries in Time*. Cambridge, MA: MIT Press.

Hughes, R. E. (1983). From ignose to hexuronic acid to vitamin C. *Trends in Biochemical Sciences*, 8, 146–7.

Hunt, B. (1991). *The Maxwellians*. Ithaca, NY: Cornell University Press.

Huygens, C. (1690/1997). *Discourse on the Cause of Gravity*, trans. K. Bailey. Mimeographed.

Illari, P. & Williamson, J. (2010). Function and organization: comparing the mechanisms of protein synthesis and natural selection. *Studies in History and Philosophy of the Biological and Biomedical Sciences*, 41, 279–91.

(2011). Mechanisms are real and local. In P. Illari, F. Russo and J. Williamson, eds., *Causality in the Sciences*. Oxford: Oxford University Press, pp. 818–44.

(2012). What is a mechanism? Thinking about mechanisms across the sciences. *European Journal of Philosophy of Science*, 2, 119–35.

(2013). In defense of activities. *Journal for General Philosophy of Science*, 44, 69–83.

Ioannidis, S. & Psillos, S. (2017). In defense of methodological mechanism: the case of apoptosis. *Axiomathes*, 27, 601–19.

Kaiser, M. I. (2018). The components and boundaries of mechanisms. In S. Glennan and P. Illari, eds., *The Routledge Handbook of Mechanisms and Mechanical Philosophy*. New York: Routledge, pp. 116–30.

Kaiser, M. & Craver, C. F. (2013). Mechanisms and laws: clarifying the debate. In H. K. Chao, S. T. Chen and R. L. Millstein, eds., *Mechanism and Causality in Biology and Economics*. Dordrecht: Springer, pp. 125–45.

Kaiser, M. I. & Krickel, B. (2017). The metaphysics of constitutive mechanistic phenomena. *The British Journal for the Philosophy of Science*, 68, 745–79.

Kanduc, D., Mittelman, A., Serpico, R., Sinigaglia, E., Sinha, A. A., Natale, C., Santacroce, R., Di Corcia, M. G., Lucchese, A., Dini, L., Pani, P., Santacroce, S., Simone, S., Bucci, R. & Farber, E. (2002). Cell death: apoptosis versus necrosis. *International Journal of Oncology*, 21, 165–70.

Kant, I. (1790/2008). *Critique of Judgement*, trans. N. Walker and J. C. Meredith. Oxford: Oxford University Press.

Kaplan, D. M. (2012). How to demarcate the boundaries of cognition. *Biology and Philosophy*, 27, 545–70.

Kerr, J. F. R. (1971). Shrinkage necrosis: a distinct mode of cellular death. *Journal of Pathology*, 105, 13–20.

(2002). History of the events leading to the formulation of the apoptosis concept. *Toxicology*, 181–2, 471–4.

Kerr, J. F. R., Wyllie, A. H. & Currie, A. R. (1972). Apoptosis: a basic biological phenomenon with wide-ranging implications in tissue kinetics. *British Journal of Cancer*, 26, 239–57.

Kitcher, P. (1989). Explanatory unification and causal structure. In P. Kitcher and W. Salmon, eds., *Scientific Explanation*, Minnesota Studies in the Philosophy of Science, vol. 13. Minneapolis: University of Minnesota Press, pp. 410–505.

(1993). Function and design. *Midwest Studies in Philosophy*, 18, 379–97.

Klein, M. J. (1972). Mechanical explanation at the end of the 19th century. *Centaurus*, 17, 58–82.

Kluve, J. (2004). On the role of counterfactuals in inferring causal effects. *Foundations of Science*, 9, 65–101.

Kreines, J. (2004). Hegel's critique of pure mechanism and the philosophical appeal of the Logic project. *European Journal of Philosophy*, 12, 38–74.

Krickel, B. (2018). *The Mechanical World: The Metaphysical Commitments of the New Mechanistic Approach*. Cham: Springer.

Lakhani, S., Dilly, S. & Finlayson, C. (2009). *Basic Pathology: An Introduction to the Mechanisms of Disease*, 4th ed. London: Hodder Arnold.

Lange, M. (2000). *Natural Laws in Scientific Practice*. Oxford: Oxford University Press.

(2009). *Laws and Lawmakers: Science, Metaphysics, and the Laws of Nature*, New York: Oxford University Press.

Larmor, J. (1894). A dynamical theory of the electric and luminiferous medium (part I). *Philosophical Transactions of the Royal Society*, 185, 719–822 (reprinted in J. Larmor, *Mathematical and Physical Papers*, vol. 1, [Cambridge: Cambridge University Press, 1929]).

Leibniz, G. W. (1989). *G. W. Leibniz: Philosophical Essays*, ed. R. Ariew and D. Garber. Indianapolis, IN: Hackett Publishing Company.

Leuridan, B. (2010). Can mechanisms really replace laws of nature? *Philosophy of Science*, 77, 317–40.

(2012). Three problems for the mutual manipulability account of constitutive relevance in mechanisms. *The British Journal for the Philosophy of Science*, 63, 399–427.

Levin S., Bucci, T. J., Cohen, S. M., Fix, A. S., Hardisty, J. F., LeGrand, E. K., Maronpot, R. R. & Trump, B. F. (1999). The nomenclature of cell death: recommendations of an ad hoc committee of the society of toxicologic pathologists. *Toxicologic Pathology*, 27, 484–90.

Levy, A. (2013). Three kinds of New Mechanism. *Biology and Philosophy*, 28, 99–114.

Lewis, D. (1973). *Counterfactuals*. Cambridge, MA: Harvard University Press.

(1986a). *Philosophical Papers*, vol. 2. Oxford: Oxford University Press.

(1986b). Causation. In *Philosophical Papers*, vol. 2. Oxford: Oxford University Press, pp. 159–213.

Lind, J. (1753). *A Treatise of the Scurvy, in Three Parts, Containing an Inquiry into the Nature, Causes, and Cure of That Disease. Together with a Critical and Chronological View of What Has Been Published on the Subject*. Edinburgh: Sands, Murray and Cochran.

Linster, C. L. & Van Schaftingen, E. (2007). Vitamin C biosynthesis, recycling and degradation in mammals. *FEBS Journal*, 274, 1–22.

Lockshin, R. A. (2008). Early work on apoptosis, an interview with Richard Lockshin. *Cell Death and Differentiation*, 15, 1091–5.

(2016). Programmed cell death 50 (and beyond). *Cell Death and Differentiation*, 23, 10–17.

Lockshin, R. A. & Williams, C. M. (1964). Programmed cell death – II. Endocrine potentiation of the breakdown of the intersegmental muscles of silkmoths. *Journal of Insect Physiology*, 10, 643–9.

Lockshin, R. A. & Zakeri, Z. (2001). Programmed cell death and apoptosis: origins of the theory. *Nature Reviews Molecular Cell Biology*, 2, 545–50.

Machamer, P. (2004). Activities and causation: the metaphysics and epistemology of mechanisms. *International Studies in the Philosophy of Science*, 18, 27–39.

Machamer, P., Darden, L. & Craver, C. F. (2000). Thinking about mechanisms. *Philosophy of Science*, 67, 1–25.

Mackie, J. L. (1973). *Truth, Probability and Paradox*. Oxford: Clarendon Press.

(1974). *The Cement of the Universe*. Oxford: Clarendon Press.

Magiorkinis, E., Beloukas, A. & Diamantis, A. (2011). Scurvy: past, present and future. *European Journal of Internal Medicine*, 22, 147–52.

Majno, G. & Joris, I. (1995). Apoptosis, oncosis, and necrosis. An overview of cell death. *American Journal of Pathology*, 146, 3–15.

Maldonado, G. & Greenland, S. (2002). Estimating causal effects. *International Journal of Epidemiology*, 31, 422–9.

Matthews, L. J. & Tabery, J. (2018). Mechanisms and the metaphysics of causation. In S. Glennan and P. Illari, eds., *The Routledge Handbook of Mechanisms and Mechanical Philosophy*. New York: Routledge, pp. 131–43.

Mayr, E. (1959). Typological versus population thinking. In B. J. Meggers, ed., *Evolution and Anthropology: A Centennial Appraisal*. Washington, DC: Anthropological Society of Washington, pp. 409–12.

Maxwell, J. C. (1873). *A Treatise on Electricity and Magnetism*, 3rd ed., vol. 2. Oxford: Clarendon Press.

McLaughlin, P. (1990). *Kant's Critique of Teleology in Biological Explanation*. Lewiston, NY : Edwin Mellon Press.

McManus, F. (2012). Development and mechanistic explanation. *Studies in History and Philosophy of Biological and Biomedical Sciences*, 43, 532–41.

Menzies, P. (2012). The causal structure of mechanisms. *Studies in History and Philosophy of Biological and Biomedical Sciences*, 43, 796–805.

Millikan, R. G. (1984). *Language, Thought, and Other Biological Categories: New Foundations for Realism*. Cambridge, MA: MIT Press.

Mitchell, S. D. (2000). Dimensions of scientific law. *Philosophy of Science*, 67, 242–65.

(2008). Causal knowledge in evolutionary and developmental biology. *Philosophy of Science*, 75, 697–706.

Mumford, S.(2004). *Laws in Nature*. London: Routledge.

Neander, K. (1991). Functions as selected effects: the conceptual analyst's defense. *Philosophy of Science*, 58, 168–84.

Needham, J. (1943). *Time: The Refreshing River*. London: Allen and Unwin.

Nelson, D. L., Lehninger, A. L. & Cox, M. M. (2008). *Lehninger Principles of Biochemistry*, 5th ed. New York: W. H. Freeman.

Newton, I. (2004). *Philosophical Writings*, ed. A. Janiak. Cambridge: Cambridge University Press.

Nicholson, D. J. (2011). The concept of mechanism in biology. *Studies in History and Philosophy of Biological and Biomedical Sciences*, 43, 152–63.

(2019). Is the cell *really* a machine? *Journal of Theoretical Biology*, 477, 108–26.

Oppenheim, P., & Putnam, H. (1958). The unity of science as a working hypothesis. In H. Feigl, M. Scriven and G. Maxwell, eds., *Concepts, Theories, and the Mind-Body Problem*. Minneapolis: University of Minnesota Press, pp. 3–36.

Pearl, J. (2000). Comment. *Journal of the American Statistical Association*, 95, 428–31.

Pearl, J. & Mackenzie, D. (2018). *The Book of Why: The New Science of Cause and Effect*. New York: Basic Books.

Poincaré, H. (1890/1901). *Électricité et optique: la lumière et les théories électromagnétiques*, 2nd ed. Paris: Gathier-Villairs.

(1897). Les idées de Hertz sur la mécanique. *Revue Générale des Sciences*, 8, 734–43 (reprinted in *Ouevres de Henri Poincaré* [Paris: Gauthier-Villars, 1952], vol. 7, 231–50).

(1900). Relations entre la physique expérimentale et de la physique mathématique. *Revue Générale des Sciences*, 11, 1163–5.

(1902/1968). *La science et l'hypothése*. Paris: Flammarion.

Povich, M. & Craver, C. F. (2018). Mechanistic levels, reduction and emergence. In S. Glennan and P. Illari, eds., *The Routledge Handbook of Mechanisms and Mechanical Philosophy*. New York: Routledge, pp. 185–97.

Proskuryakov, S. Y. & Gabai, V. L. (2010). Mechanisms of tumor cell necrosis. *Current Pharmaceutical Design*, 16, 56–68.

Psillos, S. (1995). Poincaré's conception of mechanical explanation. In J.-L. Greffe, G. Heinzmann and K. Lorenz, eds., *Henri Poincaré: Science and Philosophy*. Berlin: Academie Verlag; Paris: Albert Blanchard.

(1999). *Scientific Realism: How Science Tracks Truth*. London: Routledge.

(2002). *Causation and Explanation*. Chesham Acumen; Montreal: McGill-Queens University Press.

(2004). A glimpse of the secret connexion: harmonising mechanisms with counterfactuals. *Perspectives on Science*, 12, 288–319.

(2007). Causal explanation and manipulation. In J. Person and P. Ylikoski, eds., *Rethinking Explanation*, Boston Studies in the Philosophy of Science, vol. 252. Dordrecht: Springer, pp. 97–112.

(2014). Regularities, natural patterns and laws of nature. *Theoria*, 79, 9–27.

Ramsey, F. P. (1925). Universals. *Mind*, 34, 401–17.

Rang, H. P., Ritter, J. M., Flower, R. J. & Henderson, G. (2016). *Rang & Dale's Pharmacology*, 8th ed. London: Elsevier Churchill Livingstone.

Reichenbach, H. (1956). *The Direction of Time*. Berkeley: University of California Press.

Ridley, M. (2004). *Evolution*, 3rd ed. Malden, MA: Blackwell Publishing.

Romero, F. (2015). Why there isn't interlevel causation in mechanisms. *Synthese*, 192, 3731–55.

Ross, L. N. (2021). Causal concepts in biology: how pathways differ from mechanisms and why it matters. *The British Journal for the Philosophy of Science*, 72, 131–58.

Rubin, D. B. (1978). Bayesian inference for causal effects: the role of randomization. *The Annals of Statistics*, 6, 34–58.

Ruiz-Mirazo, K., Peretó, J. & Moreno, A. (2004). A universal definition of life: autonomy and open-ended evolution. *Origins of Life and Evolution of the Biosphere* 34, 323–46.

Russell, B. (1905). On denoting. *Mind*, 14, 479–93.

Russo, F. & Williamson, J. (2007). Interpreting causality in the health sciences. *International Studies in the Philosophy of Science*, 21, 157–70.

Salmon, W. (1984). *Scientific Explanation and the Causal Structure of the World*. Princeton, NJ: Princeton University Press.

(1997). Causality and explanation: a reply to two critiques. *Philosophy of Science*, 64, 461–77.

Saunders, J. W., Jr. (1966). Death in embryonic systems. *Science*, 154, 604–12.

Schiemann, G. (2008). *Hermann von Helmholtz's Mechanism: The Loss of Certainty*. Berlin: Springer.

Schofield, R. E. (1970). *Mechanism and Materialism: British Natural Philosophy in An Age of Reason*. Princeton, NJ: Princeton University Press.

Schultz, J. (2002). *The Discovery of Vitamin C by Albert Szent-Gyögyi*. American Chemical Society National Historic Chemical Landmarks. www.acs.org/content/acs/en/education/whatischemistry/landmarks/szentgyorgyi.html (accessed February 11, 2019).

Sellars, W. (1958). Counterfactuals, dispositions, and the causal modalities. In H. Feigl, M. Scriven and G. Maxwell, eds., *Concepts, Theories, and the Mind-Body Problem*. Minneapolis: University of Minnesota Press, pp. 225–308.

Shiozaki, E. N. & Shi, Y. (2004). Caspases, IAPs and Smac/DIABLO: mechanisms from structural biology. *Trends in Biochemical Sciences*, 39, 486–94.

Simon, H. A. & Rescher, N. (1966). Cause and counterfactual. *Philosophy of Science*, 33, 323–40.

Skipper, R. & Millstein, R. (2005). Thinking about evolutionary mechanisms: natural selection. *Studies in History and Philosophy of Biological and Biomedical Sciences*, 36, 327–47.

Slack, J. M. W. (2005). *Essential Developmental Biology*, 2nd ed. Malden, MA: Blackwell Publishing.

Sloviter, R. S. (2002). Apoptosis: a guide for the perplexed. *Trends in Pharmacological Sciences*, 23, 19–24.

Sober, E. (1984). *The Nature of Selection*. Chicago: University of Chicago Press.

Stalnaker, R. (1968). A theory of conditionals. In N. Rescher, ed., *Studies in Logical Theory*. Oxford: Blackwell, pp. 98–112.

Stone, R. (1993). The assumptions on which causal inferences rest. *Journal of the Royal Statistical Society B*, 55, 455–66.

Thagard, P. (1999). *How Scientists Explain Disease*. Princeton, NJ: Princeton University Press.

Todd, W. (1964). Counterfactual conditionals and the presuppositions of induction. *Philosophy of Science*, 31, 101–10.

Tooley, M. (1977). The nature of laws. *Canadian Journal of Philosophy*, 7, 667–98.

Tortora, G. J. & Derrickson, B. (2012). *Principles of Anatomy & Physiology*, 13th ed. Hoboken, NJ: John Wiley & Sons.

von Dassow, G. & Munro, E. (1999). Modularity in animal development and evolution: elements of a conceptual framework for evo-devo. *Journal of Experimental Zoology*, 285, 307–25.

Walsh, D. M. (2006). Organisms as natural purposes: the contemporary evolutionary perspective. *Studies in History and Philosophy of Biological and Biomedical Sciences*, 37, 771–91.

Waskan, J. (2011). Mechanistic explanation at the limit. *Synthese*, 183, 389–408.

Waters, C. K. (1998). Causal regularities in the biological world of contingent distributions. *Biology and Philosophy*, 13, 5–36.

Weinberg, J. (1951). Contrary-to-fact conditionals. *Journal of Philosophy*, 48, 17–22.

Williamson, J. (2011). Mechanistic theories of causality part II. *Philosophy Compass*, 6, 433–44.

Williamson, J. & Wilde, M. (2016). Evidence and epistemic causality. In W. Wiedermann and A. von Eye, eds., *Statistics and Causality: Methods for Applied Empirical Research*. Hoboken, NJ: Wiley and Sons, pp. 31–41.

Wilson, M. D. (1999). *Ideas and Mechanism: Essays on Early Modern Philosophy*. Princeton, NJ: Princeton University Press.

Wimsatt, W. C. (1976). Reductionism, levels of organization, and the mind–body problem. In G. Globus, I. Savodnik and G. Maxwell, eds., *Consciousness and the Brain*. New York: Plenum Press, pp. 199–267.

Wolpert, L., Beddington, R., Jessell, T., Lawrence, P., Meyerowitz, E. & Smith, J. (2002). *Principles of Development*, 2nd ed. Oxford: Oxford University Press.

Wong, R. S. Y. (2011). Apoptosis in cancer: from pathogenesis to treatment. *Journal of Experimental and Clinical Cancer Research*, 30, 87.

Woodger, J. H. (1929). *Biological Principles: A Critical Study*, London: Routledge & Kegan Paul.

Woodward, J. (1997). Explanation, invariance and intervention. *Philosophy of Science*, 64, S26–S41.

(2000). Explanation and invariance in the special sciences. *British Journal for the Philosophy of Science*, 51, 197–254.

(2002). What is a mechanism? A counterfactual account. *Philosophy of Science*, 69, S366–S377.

(2003a). *Making Things Happen: A Theory of Causal Explanation*. New York: Oxford University Press.

(2003b). Counterfactuals and causal explanation. *International Studies in the Philosophy of Science*, 18, 41–72.

(2011). Mechanisms revisited. *Synthese*, 183, 409–27.

(2013). Mechanistic explanation: its scope and limits. *Aristotelian Society Supplementary Volume*, 87, 39–65.

(2014). A functional account of causation; or, a defense of the legitimacy of causal thinking by reference to the only standard that matters – usefulness (as opposed to metaphysics or agreement with intuitive judgment). *Philosophy of Science*, 81, 691–713.

(2015). Interventionism and causal exclusion. *Philosophy and Phenomenological Research*, 91, 303–47.

(2020). Levels: What are they and what are they good for?. In K. S. Kendler, J. Parnas and P. Zachar, eds., *Levels of Analysis in Psychopathology: Cross Disciplinary Perspectives*. Cambridge: Cambridge University Press, pp. 424–49.

Index

Printed in the United States
by Baker & Taylor Publisher Services